COVID-19 and the Environment

COVID-19 and the Environment

Edited by:
Austin Mardon
Catherine Mardon

Written by:
Ching Yuan
Chitrini Tandron
Tenzin Yehsopa

2020

A Golden Meteorite Press Book.
Printed in Canada.

© 2020 copyright in Canada by Austin Mardon. All rights reserved. This book or any portion thereof may not be reproduced or used in any manner whatsoever without the express written permission of the publisher except for the use of brief quotations in a book review or scholarly journal.

First Printing: 2020

Typeset and Cover Design by Fariha Khan

Email: aamardon@yahoo.ca
Telephone: 1-(587)-783-0059
Website: www.goldenmeteoritepress.com

Additional copies can be ordered from:
Suite 103 11919 82 Street NW
Edmonton, AB
T5B 2W4
CANADA

ISBN 978-1-77369-161-9

We acknowledge the support of Canada Service Corps, TakingITGlobal, and the Government of Canada in promotional materials associated with the Project.

Chapter 1

Introduction

The environment, as far as humanity is concerned, consists of two parts: the natural biosphere and the man-made surroundings that most people live in. These two components are inexplicably linked with each other as well as with humanity itself. Changes to the human population will initiate changes in the environment, and vice versa. This web of relations is often so complex that it's near impossible to tell which is the cause and which event is the effect.

The coronavirus, for example, is a zoonotic virus that jumped from animal to human, wreaking havoc on our modern-day society. The way humanity has reacted to the virus then turns around and affects our immediate surroundings. In this way, COVID-19 affects the environment. However this relationship also works the other way around: the environment too, has profound effects on the infection rate of COVID-19. Humanity is caught in the middle of this web, and even now, half a year after the epidemic went global, the general public still has only vague ideas about how this interconnectivity functions

Our book seeks to define this confusing pattern of relations from three perspectives: how the virus is affecting human populations to change the environment, how the environment impacts COVID-19 spread, and how COVID-19 is impacting the environment of the future.

The Coronavirus is Changing the Environment

Perhaps the most important issue about COVID-19 and the environment is the question of how COVID-19 is affecting the present environment. As people all over the world settle into quarantine, traffic levels decrease and non-essential businesses close temporarily. In addition, lockdown will decrease overcrowding and the concentration of human activity, particularly in urban areas. The decrease in human activity is changing the very air we breathe as CO2 and other emission levels go down. Decrease in travel will also play quite a large role in changing air quality and emission of levels of air pollutants. In short, COVID-19's impact on the human population's lifestyle choices will cause changes in every aspect of the environment that human activity is tied to. This includes not only air pollution levels, but also water quality and wildlife biodiversity.

Many are under the impression that decreased human activity outside automatically means a betterment of the natural environment. However, there are aspects of COVID-19's negative effect on the environment that also must be discussed. Obvious negative changes for the environment include increased production of masks and medical waste, as well as higher levels of water and soap consumption.

All of the aforementioned changes in the environment, both good and bad, have profound effects on human health. Needless to say, changes to air quality, water consumption, and other environmental effects brought about by COVID-19 will directly impact human health both in the present as well as far into the future.

How COVID-19 is Affecting our Future Environment

The memory of the fear associated with the virus, as well as the habits that people picked up during the shift to quarantine will linger long after the pandemic abates. Humanity's outlook of the future is rapidly changing as the pandemic continues, and there seems to be more good than bad to this difference.

Firstly, the fear of infection is already motivating people to increase the development and usage of automated equipment. These technologies were already on the market before the virus, but the urgency of the situation brought on by the pandemic has pushed these previously marginalized technologies into the forefront of society's focus. Increased use of technologies foster greater adaptations and developments, and indeed much of these technologies are already seeing major positive changes as they expand to meet the growing demand. In short, there is now a highly viable market in developing automated equipment due to rising demands from health-conscious customers. Once people begin to use automated equipment and see their efficiency, it is likely that this increased usage will continue into the future after the pandemic. A lot of the reasons why the automated technologies weren't in popular demand before the pandemic was because people had only limited knowledge of them. However, the global situation is forcing humanity's hand, acting as a pressuring incentive for people to expand their knowledge and use of automated technology. In other words, the pandemic is a highly effective introduction to automated technology, and people are more likely than not to continue using automated technology once they become accustomed to it.

Secondly, many of these changes in present modes of thinking lead to major paradigm shifts in fields such as architecture and urban design. The coronavirus has made people realize the important role that infrastructure and architecture plays in healthy living. Before the pandemic, health issues were considered by the average layman from an almost strictly biological perspective. But now the general public is becoming more and more aware of the modes of infection, as well as the role that public and private spaces play in spreading illnesses. This conscientiousness will no doubt influence future demand for home and public space design, indicating lasting change affecting generations to come.

The Pandemic is Environment Dependent

Aspects of the environment such as temperature, humidity, and UV ray exposure are all factors that have been shown to affect the half-life of airborne viruses like the coronavirus. A shorter half-life can significantly reduce the infection numbers - that is, if the environment is able to impact COVID-19 in the first place. Whether or not the environment plays a big role in the life cycle of COVID-19 brings to mind another question: is COVID-19 a seasonal virus, or is it capable of becoming a synchronized infection spanning the entire globe? It's important not only for scientists, but also for the general public to know these connections between COVID-19 and the environment.

Besides factors from the natural environment, the man-made geopolitical environment also may play a role in the severity of the COVID-19 pandemic. Developing countries and first world countries are reacting to the virus very differently. This is because many characteristics of developing nations puts them in a poor position to handle the epidemic once it expands into their borders. Understanding the various geopolitical factors that can impact health issues is vital to the process of adapting approaches to solve the issues with minimal collateral damage.

Throughout the book, many of these topics will be explored in depth to yield a greater understanding of the interconnectivity between COVID-19 and the environment, as well as the role that humans play within this interaction. The book serves to reduce confusion by presenting summarized information based on various different sources. Throughout the course of the book, we hope to spread more awareness about the way that epidemics and the environment mutually affect each other, as well as dispel common misconceptions about this interaction.

Chapter 2

COVID-19 and Climate Change

The current pandemic has really taken the whole world by storm. Countries around the world have begun discussing what the future may hold and the overall lasting impact this pandemic will have on the earth. As we look at the Earth's history, we can tell that humans have been attributed to most of the damaging effects. Over the years, the amount of scientific research done on combating global emissions have risen astronomically. More and more nations are agreeing on ways to combat this global issue. Now with this new global issue, there needs to be substantial change in order to mitigate the effects of CO2 in our atmosphere. During these trying times of the COVID-19 outbreak, there has been rising concern of what will happen to our surrounding environment and the ongoing issue of global warming. Though the outcomes of the pandemic have been more than horrendous, there have been some positive outcomes in regard to CO2 emissions. As the world becomes subsumed to disaster and outbreak, the CO2 emissions and the very ozone layer effects we once worried about have shown some surprising turn of events.

The research of the ongoing COVID-19 global pandemic has just begun. As this spreads, we can understand the changes it has on not only human population and migration but also the factors on climate change. This topic will encompass the ozone layer and its effects on COVID-19 pandemic. The ozone is a layer that is helping to hold in the heat from natural sun's radiation like the other greenhouse gases such as carbon dioxide, methane, nitrous oxide and other industrial gases.[5] Greenhouse gases trap heat from the sun in Earth's atmosphere which has kept the climate habitable for humans and millions of other species. Those gases over the last couple of decades have been out of balance due to ongoing human activity which is causing climate change and other harmful secondary effects. Climatologists throughout the decades have been warning governments and institutions about the excessive amount of greenhouse gases in our atmosphere that is trapping too much heat which is slowly making our global average temperature increase. The risk of the ozone not only at the planetary level but also can have increased levels of UV radiation that will reach earth and harm those on it.

Decreased CO2 Emissions

It only took a few weeks before the once polluted cities in major urban areas like Shanghai and New Delhi became clear of a large amount of smog from pollution in the air. The once air polluted areas were now displaying skies and rays of light that weren't visible before. In India there were photos circulating of the Himalayas which were now visible. They were obscured by heavy smog and pollution before. This is a familiar type of event that happened in many parts of the world. Nature began to feel like it was restoring itself. Before the pandemic outbreak, the emissions of carbon dioxide were rising world-wide. Studies show that they were rising 1% per year over the previous decade.[3] This change in CO2 in our atmosphere seems to be very minimal but in actuality has any drastic effects on the natural surroundings. Carbon dioxide in itself increases temperatures, can extend the growing season and can increase humidity levels in certain areas of the world. All of these effects which were just mentioned also have a domino effect causing a number of secondary effects such as a fluctuation in soil moisture levels and stress on plants. A study published in the Journal Nature Climate Change shows that the daily emissions have dramatically decreased, by 17% or 17 million tonnes of carbon dioxide.[6] This occurred globally during the peak of confinement and quarantine measures in April. These numbers were compared to the mean daily levels that were taken in 2019 dropping to levels that were observed during 2006. Despite this, when looking at carbon emissions as a collective and not at an individual level, the concentration of carbon dioxide in the atmosphere was the highest ever recorded in human history in May 2020.[3]

The emissions from day to day surface transport such as cars, railways and other forms of transportation accounted for almost half of the decrease in global emissions during peak confinement.[3] Emissions from industrial plants and from generators powering a multitude of cities account for further 43% decrease in daily global emissions. Other pollutant industries such as aviation were shut down during lockdown but only accounted for 3% of global emissions, therefore had a 10% decrease in emission during the pandemic.[3] In this study they looked at individual countries and saw that on average 26% of the emissions decreased at the peak of their confinement.[3] There is clear indication that society's response to the global pandemic has created a substantial impact on our CO2 emission levels. These numbers seem to be temporary as our social response happened due to the pandemic, they don't reflect any sort of structural or institutional change.

This is certainly something global leaders can learn from. As scientists come out with more and more research, there are opportunities to adapt and learn these sustainable strategies to implement in our lives to further help and mitigate our ongoing carbon dioxide emissions issue. The estimated total change in emissions alone from the pandemic account to million tonnes of CO2. The largest change was seen in China where there are known to be high levels of CO2 in the air. A temporary reduction in daily global CO2 emissions during COVID-19 was merely caused by forced confinement, now if the government remains committed to creating systemic change, we can strive towards long term goals.

Ozone Layer Effects on COVID-19

The ozone is essentially the region of the upper atmosphere that contains high concentrations of ozone molecules. Its primary role is to effectively block almost all solar radiation of certain wavelengths from reaching Earth's surface including UV and other forms of radiation that can kill or harm living beings on earth.[7] Back in the 1980's ozone depletion was a hot topic and the issue arose by many leading scientists. The ozone depletion they were examining at the time was causing the southern air currents to be driven further south, subsequently causing major climatic changes all across the globe but primarily in a few concentrated areas. These changes include differing rainfall patterns and ocean currents. These effects had lasting impacts on ocean currents and salinity in places like South America, East Africa, and Australia.[4]

More recently, there are studies being done that claim that the ozone hole they were concerned about in the 80's have just closed weeks after it formed over the Arctic circle. While pollution levels are continuing to go down worldwide amid the pandemic, they deem it's unlikely that it's due to the global lockdown. Scientists believe this hole was not caused by air quality changes or human activity but by a strong Arctic polar vortex.[1] It is common to have ozone holes developing over the Arctic and Antarctic every year especially during the spring months. It seems that there is a false narrative circulating around the internet that the ozone has started regenerating and healing due to worldwide lockdown procedures. Many images including wild animals trotting around, dolphins in Venice canals and many others that have been circulating through various social media platforms to present this idea in a more digestible fashion. Unfortunately, this is not the case.

The article published by Nature tries to elaborate that the pause in circulation trends is merely owed to internal or natural variability of the climatic system. It talks about the Montreal Protocol which is a treaty signed by various countries to minimize the use of ozone depleting substances. These include refrigerator and air conditioner emission's, industrial solvents and more. They are convinced that the restoration of the ozone layer is due to the Montreal Protocol due to reported numbers of less ozone-depleting substances since the year 2000.[2] They believe since it has been three decades after 197 countries signed the Montreal Protocol in 1987, we are now reaping the benefits. The issue that scientists point out is that ozone-depleting substances have long shelf lives and this means that a full recovery of the ozone layer and the subsequent holes won't happen for several decades. This isn't entirely in itself connected to the COVID-19 outbreak. This is believed to be merely due to environmental impacts as a result of worldwide changes and decrease in emissions since this protocol.

Due to the significant decline in production and industrial processes the level of air pollution throughout the globe has drastically decreased. Services and various companies reduced their scales of operations including mass productions. Some have stopped all operations at one point during the quarantine period. Some have argued that this is not due to COVID-19 at all but merely a coincidence due to the climate zone or particular season. There are current studies being done on the extent of which the lockdown has had a role in the recovery of the ozone. If this trend continues of having a large decline in mass production and services being provided, the ozone recovery story may help uncover more information. Even though the lock down is surely showing some promising signs of nature resorting itself, the restoration of the ozone layer is primarily through the Montreal Protocol.

Conclusion

In conclusion the climate data does indicate that COVID-19 could end up following similar trends of outbreaks in the past such as SARS. There is more research to be done as there are some findings that are deemed to be contradictory. As emissions in individual nations are decreasing, we also received the highest ever recorded carbon dioxide in our atmosphere. As of now all the research is still under extensive review since this outbreak is fairly recent. Many various aspects of the data that was covered in this topic can be contradicting, but this doesn't mean that one is correct and the others are not. The resulting differences could be a result of different procedures or looking at different populations. While scientists are still drawing information on the relationship between climate change and CO2, it's important to not only consider the climate conditions but also the important social changes taken place to help mitigate the effects of COVID-19.

Chapter 3

Air and Water Quality

The spread of COVID-19 in this short period of time has brought forth a dramatic decrease in industrial activities and mass production. These restrictions on processes that would normally cause great harm to surrounding environments have been a blessing for nature. There are reports from countries worldwide that indicate environmental conditions including air and water quality are improving. As emissions continue to decrease the quality of air in even the busiest of cities have shown an unprecedented improvement. With the help of remote sensing images, researchers are able to map the areas that had a significant decline in air pollution after lockdowns. What is important to highlight is that factors influencing pollution in the air also have an effect on the water conditions on the surface. The atmosphere and surface water quality are highly connected. The systems are essentially integrally linked. This raises other concerns such as viral diseases being spread through particulate matter and atmospheric pollutants being carried to our groundwater systems.

Air Quality

Air pollution is a highly important topic for many researchers across the world because it has a big toll on human health in populations. The particular importance is surrounding the studies that show that air pollution is linked to influenza and other viral diseases. There is more research being done on the respiratory tract and certain infections that can be caused by severe air pollution, particularly high nitrogen dioxide content. The particulate matter which are suspended in the air can be hazardous and can remain in the air for a prolonged period of time. If viral particles, in particular, are deeply inhaled within the respiratory tract, the virus can potentially penetrate the lung's epithelial cells and move to other organs and vital parts of the body causing an infection.[2] A study that was done during the SARS epidemic in China in 2003 found that mortality rates were higher in urban regions with high levels of ambient air pollution compared to low pollution areas.[2] This was measured by the Air Pollution Index (API). However, it is important to note that studies show both household and ambient air pollution have led to risks in certain repository diseases and is not just one of these factors but a combination. The risks of COVID-19 have been claimed to be transmitted more rapidly between people living in close proximity such as centralized urban communities with high densities.

The Centre for Research on Energy and Clean Air reported that methods to contain the spread of the coronavirus such as mandatory quarantine for individuals and travel restrictions has resulted in the outcome of a 25% reduction of carbon emissions just in China.[1] What they did in this study was compared China to the two different time periods. China produced 200 million fewer metric tons of carbon dioxide than the same period in 2019.[1] This is due to a multitude of factors but some important ones include reduction of air traffic, ground traffic, oil refinery, industries and coal consumption.[1] Now when comparing this back to the health effects we can already see that this lockdown would have essentially saved many lives.

NASA has also been doing its own work on air quality since COVID-19 outbreak occurred. They monitored the NO_2 gases during the initial phase of the pandemic. They noticed that the abundance of NO_2 gases significantly decreased during the initial phase of the Pandemic particularly in China.[3] In order to monitor this, NASA essentially used technology such as various monitoring instruments to analyze and observe the ozone layer and pollutant levels such as NO_2 and various types of harmful aerosols. According to the data collected by NASA, the NO_2 pollution began in Wuhan, China and spread like wildfire to other parts of the world.[3] What is interesting is that although the pollutants of NO_2 made a significant drop during the time of lockdown, other pollutants such as aerosols remained, and therefore China was not able to achieve an air quality standard that met the criteria of health authorities despite improving drastically in some areas.

On the other hand, air pollution and COVID-19 are portrayed much differently in more recent analysis and findings. They show that these two factors may be interacting rather than being looked at as just separate entities. The Institute of Labor Economics researchers have found the areas within the Netherlands that have higher air pollution also resulted in greater numbers of COVID-19 cases and deaths.[6] At the time of this study there were more than 50,000 cases recorded of COVID-19.[6] They compared air quality readings from various municipalities that included data of various levels of pollutants discussed previously. The polluted areas with finer particulate matter in the air corresponded to more COVID-19 related deaths.[6] There were similar studies done in the United States which resulted in similar results. At this time, it is hard to completely link air pollution and COVID-19 directly, but there are more studies emerging on this topic as it may pertain a lot of answers to the pandemic as a whole.

Water Quality

What is happening in our air also has impacts on the other Earth Systems such as water quality and health. The pandemic has resulted in a dramatic decrease in atmospheric nitrogen pollutants worldwide yet the impacts on deposited nitrogen are still being studied. The effects this has on aquatic ecology remains unknown. In general, the overabundance of nitrogen in our waters and other harmful chemicals can cause excessive algae growth. When these plants decompose, the processes consume an enormous amount of oxygen and the water can be left depleted of oxygen that is needed to sustain life, thus affecting the various aquatic life in the waters. Inherently water systems are quite complex. Water quality improvements can take time to show up and carry by location. Water in a sense needs to infiltrate the ground, which in itself filters out most of the pollution. If the pollutants have the ability to reach the groundwater system, it can stay there for months before it's swept back into lakes and rivers and the rest of the hydrological cycle. Due to the complexity of water systems and the outlook of potential water improvements for the time being due to the level of urbanization occurring, many scientists believe that water improvements are going to be only for the time being. The improvements are more likely going to be at local scales.

There have been varying results of this lockdown in different countries. The most immediate results are obviously on a local scale. In India for example, there was a strict 21-day lockdown across the nation. Many industries and offices closed due to the lockdown, restricting industrial pollutants and waste. The Yamuna river is known for its tons of extreme toxins and effluents that are discharged freely into it. This type of system is not uncommon throughout India. It is estimated that every day almost 40 million litres of waste and sewage enters river bodies and only 37% of that is adequately treated.[5] The nationwide lockdown imposed on March 25 has suggested signs of improvement in water quality. Since all the major polluting industries were closed, the toxic load was lifted temporarily from the river. The issue that arises is how long will these effects last and will things resort back to normal after lockdown or will the lack of industries at the moment exacerbate these effects further? With the help of remote sensing technology to produce high quality images, scientists were able to quantitatively demonstrate the improvement in surface water quality in terms of suspended particulate matter (SPM) in Vembanad Lake, which is the longest freshwater lake in India.[7] The SPM concentration during the lockdown period decreased by 15.9% on average compared to the period before lockdown.[7] These findings also draw further attention to what will happen after the lockdown period. Due to the pollutants decreasing considerably because of industries being suspended for the time being, there needs to be considerable action done to keep this trend moving in this direction even after COVID-19.

Although there have been positive impacts there are still growing concerns over the health of our waters on a long-term scale. Due to the pandemic there is concern over the use of single use plastics and how it has merged itself into our day to day lives. There is currently not that much data or studies being done on this issue, but it continues to be a growing concern. The problem of microplastics in our oceans has been an ongoing concern since the start of climate change and when that became an issue. Microplastics are an ongoing issue because these tiny particles can easily pass through water filtration and end up in big bodies of water leaving a threat to aquatic life.[4] Single-use plastics without a doubt have been an important and key tool to fight against this pandemic, especially those who work on the front lines. The concern arises when there are stacks of plastic and medical waste piling up and floating in coastal waters. If countries are not careful on their short-term goals during this pandemic, it could set us back on our environmental goals and even public-health and safety in the future.

Conclusion

In conclusion, looking at the impact COVID-19 has on the air and water quality alone is not enough to accurately gauge the lasting effects of this pandemic. We must look at these two factors interchangeably in order to get a better understanding of the future quality of water and air. We will have to wait to see if there are any significant long-term changes. At the moment what we can do is observe the changes we see on local scales and compare them to past data. It's important to think about both short term and long-term effects when using various tools to help fight this pandemic.

Chapter 4

Impact on Wildlife and Nature

This pandemic which brought forth tragic circumstances to human life but has also provided important insights into human and wildlife interactions. The interest lies on how wildlife and mother nature has responded to these new conditions. The lockdown measures being done by humans have not only created a lasting impact on the environment and air quality, but also for the surrounding wildlife. COVID-19 in itself is quite complex. To understand this more clearly, it's better to look at Coronavirus as a large family of viruses.[2] Some of these viruses cause cold-like illnesses while others cause illnesses in various types of animals such as cows, camels and bats.[2] There is still much to learn about the virus, but it appears that it can spread from people and animals in various situations. It is also important to realize the impact it has on wildlife as an entity. The global pandemic also had some interesting effects on nature. These effects are directly caused by the virus itself, but more so the human led impacts as a result of lockdown and reduced travel. We will begin by discussing the impacts that COVID-19 have on our natural surroundings and ecosystems.

Nature

Ecosystems in nature function like the human body. When the body is healthy like a forest with diverse species and a balance in animal relationships they are less likely to be prone to diseases and potential threats. In turn, there is a common misperception that nature is "getting a break" from human activity during the pandemic. The main issue people are trying to raise currently is that outside the main urban areas the situations become extremely problematic. In rural areas, there is an ongoing increased pressure for natural resources. There have been reports of criminal activity involving land-grabbing, deforestation and illegal mining during these times. Due to the fact that governments and institutions are focused on battling this pandemic instead of using resources for conservation purposes, it has led to numerous instances where people are illegally extracting natural resources. There are numerous areas where activities and sites are left unguarded leaving easy access and ways to get away with these types of criminal activity without any sort of government intervention. These acts are expected to only increase until economies rebound and governments are able to shift their focus again on environmental conservation.[5]

Deforestation has worsened during the start of the pandemic. The panic and urgency of the pandemic has continuously diverted the attention away from the ongoing deforestation issues. Essentially the acts of deforestation have only exacerbated rather than improve during the pandemic. The issue with deforestation is that it takes away one of the largest carbon sinks in the world. These carbon sinks act as huge atmospheric carbon recycle areas so that trees and other plants can use up a big chunk of the carbon dioxide. This is only contributing towards climate change. The results of the deforestation that are happening in major parts of the world will see the outcome in years to come. Events that are likely to occur because they are indicators of the ongoing deforestation issues are droughts, fires, and extreme weather.

Much of this loss of tropical rainforests and vegetation has been focussed around the Amazon. Over the recent years there have been policy shifts made by the Brazilian government, with 3,769 square miles of Amazonian forest cover being lost.[4] Coupled with a rapidly growing population, humans are increasing their interactions with animals. These trends differ from what is normal because wild animals found in the Amazon have been naturally contained or migrated further inland. The increased contact between humans and tropical animals that resulted from deforestation raises concerns and new risks involved with various diseases and natural ecosystems essentially being disrupted. Destruction of the forests not only destroys these natural carbon sinks but also encourages the spreads of diseases. When looking at past viruses, the Ebola virus infiltrated itself into humans from fruit bats that were attracted to fruit fields. Zika virus was spread due to mosquitos being able to find solstice in urban areas with high population densities. However, it's important to keep in mind that the rise in deforestation in Amazon existed before the emergence of COVID-19. There has been an upward trend in deforestation in this part of the world since 2012, but in more recent years have taken a sharp turn due to the new President who took office in 2019.[1]

Wildlife

Due to the decrease in human activity during quarantine, there has been more wildlife activity. The start of lockdown also initiated a drop in tourism where this had positive outcomes for certain animal populations. Coastlines and beaches dropped in visitors leaving the animals who reside there to be able to move more freely and produce offspring with ease. Similar success stories were echoed in other regions of the world. There are some studies that indicate that the decrease in human activity really brought forth these positive results of freeing up space for wild animals. There have also been numerous stories circulating on how the pandemic has impacted animal migration. There have been many unusual animals being spotted in urban cities that were not commonly found there before the pandemic. With this unfortunate increase of mining and logging, this ultimately affects the habitats of animals living in these areas. The animals are then essentially forced to be confined and live in more tight areas, coming into contact with humans more often. This can offset a bunch of risks such as transmission of diseases.

The spread of COVID-19 has really only exacerbated the effects of things that were already happening. The pandemic is essentially serving as a wakeup call to our habits and actions as mankind. One interesting topic that arises when looking at the pandemics impact on wildlife are the apes. The pandemic is not only affecting human beings, but many other species are impacted as well. The great apes share about 85% of the same DNA we as humans have, the only issue is that they don't practice social distancing and merely go wherever they please in their habitat. As of now there have been no cases of COVID-19 among apes/primates.[3] It is unfortunately known that great apes have been susceptible to human pathogens and can form respiratory illnesses similar to ones in humans. The COVID-19 transmission itself is hard to quantify as there are no cases, but we know in history it's not uncommon for outbreaks to affect both humans and great apes. Just alone in 2013, with the outbreak of the common cold, it killed 5 great apes alone in Uganda.[3] Although this is an acute threat at the moment because there have been no cases found yet, the human impacts such as logging and habitat loss due to urbanization are increasingly endangering these species. There needs to be more contact between humans and apes in order for diseases to transfer, therefore being in close contact due to deforestation and other human impacts like tourism can exacerbate these affects.

Conclusion

In conclusion it is evident that there are both positives and negative outcomes when it comes to COVID-19 pandemic and the impact it has on wildlife and nature around the world. It is important for there to be stricter government regulations to prevent long term ecosystem disasters such as the loss of species or even more natural disasters. Due to the pandemic, there was a shift in focus in order to save humans from this deadly virus. It's now evident more than ever that there has to be a shift in focus on the surrounding nature and wildlife so there are no ongoing effects even after the pandemic is over.

Chapter 5

Reduction in Travel and Reduced Usage of Fossil Fuels

It is evident that our day to day lives are going to be shaped by this global pandemic for a long period of time. It will be potentially years before normal modes of transportation are back to the norms. These types of transportation include flight, public transit and boats. Currently there are mandatory checks before one can take any sort of sky transportation. These checks include but are not limited to temperature checks and can be denied boarding if passengers have a fever. The fever restriction is strictly due to it being an indicator and common symptom of COVID-19. There are also some companies who are taking more rigorous steps to administer to every passenger prior to boarding. With these new measures taking place, it is evident that travel restrictions are here to stay for now. Now the fossil fuel industry has been in serious decline for several years. This is mostly due to the rise of renewables combined with the fall in consumption from the rise of the COVID-19 pandemic. Despite the growing demand for oil-based products, the prices of fossil fuels have been falling year to year. Now, leaders and other industries are looking towards renewable resources which have been slowly growing in demand over the past few years.

Travel Restrictions

COVID-19 is a highly infectious virus. Those who are being treated are separated into groups that are asymptomatic patients and pre-symptomatic.[1] This is a serious health issue because many of the carriers of the virus can go for weeks without knowing they have the means to spread the disease. There was an outcry initially over the Chinese Government and their inability to contain the pandemic in the first place. It is important to keep in mind that trying to contain the virus in Wuhan would have been almost near to impossible considering the density of size of the area. There needed to have been a solid universal plan for managing infectious diseases if this were the case. During the start of the virus there was lots of information missing and the virus was very much up in the air as scientists scrambled to find more information to inform the public. Long term effects were unforeseen at the time, they were merely finding ways to contain it as best as possible. COVID-19 spread rapidly around the globe after that. First there were four cases which they ruled to be viral pneumonia but they were not responding to typical treatments. This is when dozens more fell ill and now the number of people infected keeps growing day by day.4 The World Health Organization believed that each infected person was spreading the disease to three other people.[6] This was during the height of traveling season and people were going back and forth by transportation methods. Some were coming back from celebrating Lunar New Years with family and loved ones. There were essentially millions travelling in and out of Wuhan before China made travel restrictions. Since the symptoms of COVID-19 only arise weeks after contact, there were people traveling to all parts of the world while spreading this virus to other travelers along the way. This further exacerbated the spread of the virus ultimately leaving the WHO to call this a global pandemic. Once this was established, travel restrictions started to occur. Borders around the globe were closing and people were scrambling to figure out lock down plans and whether to stay in the region they were in or return to their home country. This scramble went on for a few weeks before eventually most borders were shut and people were contained in the nation of residence. After that travel was very limited and most people stayed home depending on each country's wishes.

Future of Fossil Fuels

There have been discussions about the fossil fuel industry potentially on the verge of collapsing. It is estimated that over 5,600 companies in the fossil fuel industry have taken a minimum of $3 billion in aid from the US federal government due to COVID-19.[3] These types of businesses include but are not limited to drillers, mine operators, refiners and pipeline companies in this industry.[3] The International Energy Agency has announced that the outbreak of COVID-19 would essentially wipe out demand for fossil fuels because of the drop in energy demands which are 7 times greater than the global financial crisis.[2] Due to industries and transportation at an all time low world wide, there have been less demands for this type of fuel. On the other hand, there has been a steady but gradual rise of the need for renewable energy. The world is expected to see small but substantial growths of the need for renewable energy sources to make up for the shrinking demand for electricity. The impact that COVID-19 plays in the fossil fuel industry is substantial. There has been a collapse of oil market prices which have affected us here in North America. There was notice that the travel restrictions alongside the diminishing demand, it will cause the biggest drop in oil demand in 25 years. This also includes other resources such as coal demand which is also estimated to fall.[2]

Burning fossil fuels is harmful to the environment and also for public health. The product of burning fossil fuels is sulfur dioxide which can lead to extremity respiratory problems. Particulate matter is also another big issue as this is what you get at power plants and from industries and vehicles. Renewable resources on the other hand have demonstrated resilience during these hard times. But despite this resilience at the moment it is expected to slow down as we venture further into 2020. The main factor that is causing this is due to the industrial sector.[5] Beyond electricity the type of renewable energy such as transporting biofuel production is expected to drop. Adding to these harsh difficulties the low oil demands, falling gas prices are all making these renewable technologies less competitive.

Chapter 6

Health Effects of Changes in Environment

The drastic changes in lifestyle brought about by the pandemic has led to equally dramatic changes to the environment as people shack up indoors amid quarantine. These environmental changes in turn affect humans in various ways. Factors impacted by the public response to the pandemic include air quality, waste production, deforestation, water use and biodiversity. All of these factors have important implications for human health, which will be explored in this chapter.

Air Quality

With everyone staying at home, the decrease in factory production as well as the drop in traffic has decreased the amount of air pollutant released into the atmosphere.[7] However the positive impact of this decrease in air pollution is not immediate, seeing as how the recent decrease in pollutants is minimal compared to the amount of pollution already in the air. Decrease in air pollution doesn't mean what is there just goes away, greenhouse gases in the atmosphere are still at large, producing record temperatures yet again this year.[7] In addition, the economy cannot stay in this stalled state forever, and reopening would place air pollution levels right back where it was, potentially even higher as the economy scrambles to produce in order to catch up to what it has missed during the closedown.[7] With that in mind, there are still some positive postulates about the potential health benefits from an increase in air quality.

It goes without saying that air pollution has an effect on overall lung health. However, a less obvious effect of air pollution is its negative impact on the circulatory system.[2] Different components of air pollution act differently within the human body. For example, nitrogen dioxide from traffic and sulfur dioxide from burning fossil fuels can reduce lung function and exacerbate asthma.[5] Out of all the components of air pollutants, by far the worst for human health is particulate matter, some of which are small enough to enter the bloodstream to increase the risk of cardiovascular illnesses.[5] The majority of particulate matter in urban areas is generated from the burning of fossil fuels by power plants, industrial facilities, and vehicles.[5] All of these sources of particulate matter have seen a drastic decrease in their use during the quarantine, leading to postulation of a possible decrease in air pollution due to inactivity of pollution sources. In sum, it is evident that air pollution has direct effect on human health, and thus a decrease in pollutant release levels during quarantine can be assumed to have a reductionary effect on the prevalence of many illnesses.

This positive effect will likely be most visible within young children and older adults, who are sensitive to the effects of air pollution. Young children not only breathe in more air per body mass than people of other age groups, but they also have a still-developing respiratory system and immune system.[2] The combined effect of the aforementioned factors makes this age group particularly vulnerable to changes in air quality. As well, older adults have a weaker immune system and may have undiagnosed cardiovascular issues that can worsen under the effects of air pollution.[5] Because of their vulnerability to air pollution, it can be assumed that young children and older adults are also very sensitive to positive changes to air quality. Thus, the decrease in air pollutant emission during quarantine would have the most obvious impact on young children and older adults.

Increased Water Use

The usage of water has seen a worldwide increase as people become more aware and concerned about modes of infection.[7] Although on an individual household's scale, this increase in water usage does not seem to be a large change, however multiplying this slight increase by the couple billion households all over the world, this change in water use then becomes a significant statistic with many implications.[7]

It is a well-known fact that 75% of the Earth's surface is taken up by water. However, only 1% of all that water is freshwater available for human consumption. While it's true that water in the atmosphere can be collected and the salty sea water can be filtered, both processes are complicated, costly, and economically unfeasible. Therefore, the human-usable portion of the water cycle consists only of the places near or on the surface of the ground where rainwater and snowmelt collect. Because of this restriction, the entire human population is finding water to be an increasingly scarce resource as demand and use of water far exceed nature's rate of freshwater production.

Water is not destroyed when households use it for various purposes. Instead, it is contaminated and converted into an unusable part of the water cycle.[1] The term 'wasting water' doesn't mean that water is somehow disappearing after it's used, rather it refers to the conversion of usable water into unusable parts of the water cycle, which then decreases the overall supply of usable water.[1] For example, if nature was able to recycle dirtied water back into usable water faster than human civilization is using up freshwater, then there would be no water waste.

Counter to the anecdote, in reality nature's water cycle does not produce freshwater at a faster rate than the rate at which it is being used up.[1] This phenomenon is especially visible in many African countries as well as developing countries like India - countries that have a lot of population but not enough natural water resources to keep up with the demands of its people.[1]

Water filtration is an energy intensive process that also takes up a lot of time and money.[3] From another perspective, wasting water is equivalent to wasting the energy-intensive process of filtration. In other words, the resources used up in the steps of this filtration process are being wasted.[3] These resources include non-renewable fossil fuels that add to the current problems with air pollution and climate change when burned.[3]

Thus, the impacts of wasting water has complex, multifactorial effects on human health. On one hand, wasted water must be treated, using up a lot of resources in the process. The treatment process not only depletes resources from other potential uses, but also releases pollutants that could be detrimental to human health. On the other hand, drawing increased amounts of water out of ground aquifers is a highly unsustainable process.[1] Thus the sudden increase in water usage during the epidemic could have a negative effect on the long term availability of water resources all over the world, which puts countries with already scarce water resources at risk.[1]

But perhaps the worst effect of overusing water during an epidemic is the fact that it decreases the availability of water for poorer households.[6] This is especially evident in developing countries that already have wealth disparities in water access before the epidemic. When the poor have less access to water, they are incapable of performing basic sanitary and hygiene actions.[6] This could put them at further risk of contracting the deadly virus. In this way an increase in water use sustains the circulation of COVID-19 within a population, which in turn sustains the increase in water use in a ceaseless feedback loop.[6]

Wildlife and Biodiversity

The decrease in human activity during quarantine has given wildlife more breathing space. During the quarantine, the drop in tourism rates has had positive effects on sea turtle populations as they were able to lay eggs on beaches without interference from tourists.[7] Idle fishing boats meant that at least some of the fish population has been allowed a moment's respite, and the decrease in road traffic has also decreased roadkill rates significantly.[7] Perhaps the best effect of the Coronavirus is its effect on the use of animal parts in traditional Chinese medicine.[7] Due to the Coronavirus scare, there has been assumptions made about a possible decrease in demand for medicinal ingredients involving animal parts. This is good news for endangered species such as pangolins, which are often a prized ingredient in traditional Chinese medicine.

It must be noted however, that there is some fake news intermingled into the information available online. For example, the dolphins and swans in Venice canals, wild turkey in Oakland, and puma in Santiago are not sights specific to the epidemic.[7] These animals were already known to visit urban and suburban areas long before the epidemic began. However, decreased human activity does produce good results in terms of freeing up space for wild animals. Though the epidemic didn't make them appear, the lesser boat traffic in Venice did help the dolphins. Other areas where human civilization overlapped natural habitats (i.e. when roads cut off natural crossings) has seen similar sites of increased animal activity in the absence of human interference.[7]

The effects of lessened human impact on the ecosystem is not obviously linked to human health, but it is well known that biodiversity has important implications for the natural environment that human food security is entirely dependent on.[13] Crop diversity depends on the availability of wild species that breeders must use to crossbreed with domestic species.[13] On top of that, the overall ecosystem plays a large role in the crop growth itself. From providing pollination, to nutrient cycling and pest control, ecosystems and biodiversity is no doubt vital to the agriculture system that feeds the world.[13]

Humans derive medicine from natural compounds. The medicines used in daily life all had connections to nature's collection of unique chemicals.[11] From secondary plant compounds, to animal toxins, to the antibiotics produced by bacteria and fungi, nature provides humans with a host of chemicals to discover and use.[11] Loss of diversity before scientists have had the time to adequately explore the potential application of natural chemicals means that many medicines will never be discovered. Afterall, how can scientists work with an already extinct species? Thus, loss of biodiversity will directly impact human health by influencing the number of pharmaceuticals available to humans.

But how does less roadkill and more turtles link to these grand benefits of biodiversity? The answer lies in the incredible interconnectivity of nature. No one can say exactly what butterfly effect can be caused by having more turtles, but one message is very clear: the only way that humans can 'save the planet' and themselves is by increasing biodiversity wherever possible.

Deforestation

Brazil's deforestation has only worsened during quarantine as the urgency of the Coronavirus diverted attention away from the deforestation issue.[9]

This increase in deforestation is taking away one of the largest carbon sinks on the planet. Brazil is home to the Amazon rainforest, which acts as a huge atmospheric carbon recycle station as its many trees use up atmospheric carbon dioxide. Deforestation exacerbates climate change, which means more unpredictable weather changes much like what 2020 has already seen. The results of climate change have devastating effects on human health as different areas around the world each face their own climate change induced issues. Forest fires, droughts, locust swarms, and extreme rainstorms are only a taste of what is to come; and deforestation is only accelerating the process of climate change.

A more direct effect of deforestation is the increase in disease that it brings.[12] The sudden clearance of large areas of forest leaves perfect breeding ground for malaria-carrying mosquitos as shrub vegetation grows in.[12] Other factors also come into play in increasing the mosquito population, but overall one thing is clear: deforestation has a direct impact on the spread of infectious diseases that are detrimental to human health.

Increased Waste Production

Masks are commonly made from polypropylene polymer derived from petroleum oil.[8] Not only is the production process costly, the disposal of masks is also problematic. Masks are single use, and since most people aren't used to using masks, they are not very well versed on how to dispose of used masks either.[4] In addition, the lack of specialized facilities for dealing with the large amount of masks makes it even harder for people to make environmentally sustainable choices while preserving their health.10 In this manner, the masks are similar to plastic water bottles both in their widespread use as well as their environmental impact: the large amount of plastic trash is threatening the environment as more and more accumulate.[4]

The increase in mask usage paired with improper disposal measures are an immediate health threat to the waste collection workers.[10] Everyday, these workers must come into contact with improperly disposed and contaminated masks. This severely increases their chances of becoming infected. As well, they can also accidentally transfer the virus to the other households that they visit in doing their job.

A Final Word

Looking at decreases in air quality and pollution from production alone is not enough to gauge the effect of the epidemic. A more comprehensive analysis including other aspects of the environment such as wildlife and biodiversity, waste production and treatment, as well as deforestation. Overall, although there are positive effects to this epidemic, the negative implications for human health far outweigh the positive.

Chapter 7

Geoclimatic Variables and the Spread of COVID-19

Humans have attributed climate conditions to the spread of disease long before there was concrete experimental proof to back it up. One of the earliest advocates of this connection was the Greek physician Hippocrates, who defined the cause of disease as observable natural phenomena rather than invisible supernatural forces. Today, through examining historical records of outbreaks as well as collecting data on recent infectious diseases, it has been confirmed that climate conditions can exacerbate the spread of infection and affect the general progression of outbreaks.

COVID-19 is a global epidemic spanning entire continents, yet the spread of COVID-19 was most severe and concentrated in some areas more so than others. Needless to say, preventive measures and public reactions played a major role in creating these differences in trends, but there are also other less outstanding factors such as climate conditions that should not be ignored.[4] Indeed much of the scientific world has devoted their attention to the impact of climate trends on COVID-19 as well as its predecessor SARS-CoV-1 in hopes of identifying the characteristics of high risk regions in order to predict the trends in infection rates. Current research on this topic is still in its early stages but based on previous trends from other airborne viruses such as influenza, scientists have been paying special attention to the effects of humidity, temperature, and UV exposure.[4]

UV Ray Exposure

The structure of the corona virus consists of a protein envelope that encapsulates the viral DNA. UV rays from the sun are capable of disrupting DNA, which makes the UV rays especially detrimental to the virus. Hypothetically, areas with high UV exposure would have smaller infection rates since the UV rays would decrease the viability of viruses exposed to it. Indeed current trends do seem to indicate some connection between natural UV exposure and infection rates.[8] Statistical analysis revealed that locations with moderate UV exposure had a median growth rate of 26% per day while locations with high UV exposure had a median increase of 19% per day.[8] Studies show that exposure to UV may reduce the viral transmission rate, but by a very modest extent.

Temperature

One of the most common pieces of information circulating the internet is that the coming summer months would decrease the severity of the virus, as the hot temperature would interfere with the spread of the cold-loving virus. While there is some truth to this information, not all parts of it are to be taken word for word. Research and subsequent analysis only indicates a correlation between temperature and the spread of disease. This relationship is uncertain, and the connection between temperature and rate of infection is too weak to ascertain temperature's role in the spread of COVID-19.

In a study of 50 cities, areas with substantial levels of COVID-19 transmission were found to be distributed along the 30° N to 50° N latitude corridor - an area with similar weather patterns, low humidity, and mean temperatures of 5 to 11 °C.[7] As the virus spread however, infection had reached regions well outside the narrow climate band examined by the study due to travel and other human factors.[7]

While it's true that at locations with an already high temperature, further increase in temperature is linked with a decrease in the number of cases. However, research also shows that at locations with low temperature, every 1°C increase is actually associated with an increase in the number of cases.[9] These trends indicate the possibility of the virus having an optimal temperature of growth - in other words, there might be a best temperature for virus transmission, which might partially explain why the outbreak began at Wuhan during its winter season.[9] This optimal temperature is most likely in the cooler range, as predicted by the climate conditions favored by COVID-19 predecessor SARS-CoV-1.[1]

The effect of temperature on infection rates can be explained via two processes. Firstly, cooler temperatures are better for the virus because excessive heat negatively impacts the virus's stability.[5] Since COVID-19 is an airborne virus as opposed to being transmitted via direct contact or body fluids, it is more vulnerable to any changes in the temperature of the air that carries it. Secondly, cooler temperatures can also lead to an increase in COVID-19 cases due to its negative influence on the human immune system.[5] Therefore there is a higher possibility for outbreaks to occur in spring than during the rest of the year.

As it is, the survivability of airborne viruses begins to decrease only when the temperature is more than 30°C.[11] Even then, a warm climate by itself is not enough to have a large impact on the transmission of COVID-19. A Chinese study traced over 7000 cases of infection with known routes of exposure and found that only 1 out of the 7000 was known to have spread via outdoor contact.[8] Further evidence to disprove the effectiveness of temperature as an impediment to the spread of COVID-19 can be found in records of serious outbreaks in Ecuador and Brazil.[8] This goes to show that areas with tropical climates are still at large risk of wide-scale outbreak, and that temperature is only a minor factor in the spread of COVID-19.

Humidity

There is an inverse trend between humidity and viral transmission. As with the case of temperature, this correlation is mild at best and cannot compare the effect of human factors such as travel and public health measures.[11]

Under dry conditions, the particles suspended in the air are smaller. When a person sneezes, the smaller infectious aerosols can not only travel further due to their smaller and lighter size, but they can also stay suspended in the air for longer for the same reasons.[11] Having infectious aerosol particles staying suspended for longer periods is assumed to increase the likelihood of it being inhaled.[11] Through this mechanism, low humidity conditions make transmission of the virus more likely. As well, the mucus layer that lines the airway needs a constant supply of water to keep it supple. A healthy mucus layer is better at expelling viral particles than one that is dried out. Unfortunately under conditions of low humidity, the mucus does exactly that - it dries out and subsequently decreases in its capability to dispel viral particles.[11] Therefore dry conditions increase infection rates not only because it allows the viral particles to stay in the air longer, but also because it decreases the human body's immune response.

Under conditions of high humidity, the aerosols are larger and heavier. Due to this difference in size of aerosols, infectious particles are less likely to stay suspended in the air for long periods of time as they do in low humidity conditions.[11] The infectious viral particles are more likely to fall and hit surfaces, which decrease the likelihood of it being inhaled.[11]

However other studies show that under tropical conditions, overtly high humidity has been shown to actually increase the likelihood of viral infection.[11] This is because the near-condensation point humidity increases the deposition of virus-carrying particles on surfaces.[11] The high humidity also prevents evaporation of the deposited particles. Therefore under humid conditions there is an increased risk of transfer of the virus from surfaces to the human body via touch.[11] Records of a COVID-19 patient transmitting the virus to 80 people at a bath center seem to support this view as the virus was contracted under near saturation point levels of humidity.[11]

Overall, low humidity increases risk of inhaling the virus, high humidity increases likelihood of contracting the virus through touching surfaces where viral particles have deposited. This indicates a middling region of humidity level optimal for decreasing virus transmission rates. Official predictions state that humidity conditions of 40-60% may be ideal to impede COVID-19 infection.[5] Thus people living in cooler and drier regions should use humidifiers to increase humidity levels while people in warmer and very humid regions should pay special attention to sterilizing surfaces.

Conclusion

What all the climate data does indicate is that COVID-19 may end up following the trend of an asynchronous seasonal global outbreak, much like its predecessor SARS CoV-2.[1] An asynchronous, season-dependent outbreak means that winter in the Southern Hemisphere could see an increase in infection rate as the climate suitability of the Northern Hemisphere decreases in August only for the suitability of the Southern Hemisphere to increase.[10]

In this manner the worst-case scenario of a synchronous global pandemic becomes an unlikely event, but at the same time this shift of suitability could catch people off guard and create issues in international travel as people in different areas of the world are affected differently depending on the time of the year.

As for the specifics, the current research findings are often contradictory. There is no single, neat explanation of disease mechanism that can be summarized from the results. This is because most research on the novel virus is still in its stages. As of now all the research is still under the expensive process of peer review, and due to the relatively short period since the start of the pandemic is not enough for scientists to produce finalized answers to the problems that have plagued the public.[2]

Many of the results differ from each other not because some are right and others are wrong, but because different procedures as well as different criterion of analysis were used.[11] For example, some studies looked at the total number of cases while others based their analysis off of the percentage of the population infected. While neither of these data gathering methods are strictly right or wrong, the results that they produce are often very different.

Another reason that the research shouldn't be taken as word of law is that lab conditions used in testing aren't always a perfect reflection of real life.[7] For example, while it is easier to see a trend between temperature and virus viability based on tests done in a lab where everything else is controlled and only temperature varies, in real life the conditions are highly unpredictable, and a myriad of other factors are also acting to influence the virus. Another thing to consider is that a seemingly straight relationship such as the one between temperature and infection rates often have hidden dimensions behind their correlation.[11] For example, the differences in temperature may be affecting a third factor like the level of human activity, which in turn connects to infection rates. In this case it is then human activity that infection depends on, and not temperature.

On top of all this, conviction of the linkage between humidity and COVID-19 is weak, the same goes for the other two environmental factors. It must not be ignored that while the current scientific world is almost certain of environmental factors in the spread of COVID-19, there is no absolutely certain answer. Keeping an inquisitive and skeptical perspective when examining the effect of climate on COVID-19 is very important in preventing misconceptions that could put people at danger.

While there is indication of a trend between climate and COVID-19, none of the studies point to change of weather alone as enough to decrease the infection rate. In a study that compiled data from 4000 locations, the largest observed effect of the weather was a 30-40% reduction in COVID-19 transmission rate.[3] That does seem to be a lot, but even at such a substantial decrease the COVID-19 infection rate is still climbing exponentially, albeit a tiny bit slower than before.[3]

Another common misconception to avoid is that the change of season (i.e. the onset of summer) could somehow obliterate the virus. While forecasts do predict a hotter than average summer, that is not reason enough to slacken preventive measures. As mentioned before, hotter regions such as Brazil are also experiencing alarming rates of infection, and any effects due to temperature pales in comparison to public health interventions against the virus. This is because the COVID-19 virus has a half life of 1 hour, which is long enough for effective transmission, but not so long enough for environmental factors to have a significant impact on the virus.[11] It only goes to show that although the climate may impact COVID-19, at the end of the day it's still public health measures that matter the most. Taking the effect of climate into account can and do help people stay safe during the epidemic, but it should not be the case that only climate conditions are considered while ignoring the much more important social changes necessary to preserve the health of the public.

Chapter 8

Urban Development and Architectural Manmade Responses to COVID-19

Many political officials point to population density as the culprit for the high rate of COVID-19 infections. However recent studies on density, infection rate, and mortality rates indicate that there is no correlation between density and infection.[3] This analysis of data from more than 900 counties across the U.S. is enough to show that density is not a risk factor for infection.[3] This data is still consistent with earlier research that revealed a high risk of infection in crowded areas, because population density is not indicative of overcrowding. In fact, areas with high population density and good space management can have low levels of overcrowding. And as recent protests for the Anti-Mask movement has shown, there is nothing stopping people living in low density areas from forming large groups that exacerbate the spread of COVID-19. Even so, government officials speaking about social factors that contribute to infection often use the terms density and overcrowding interchangeably, adding more to the confusion between the vastly different terms. It is crowded spaces that puts people at risk of catching COVID-19, whereas density shows no correlation to infection rates.[3]

Two other concepts that are also often confused are density and population size. Epidemics spread with the movement and interactions of people, which occurs more within large populations.[3] This is mainly due to the shared transport infrastructures characteristic of large populations, where many people use the same facilities. However, this higher human activity is not related to density at all.[3] Neither is population, because large populations can live in a dispersed manner, and small populations may be constricted to high density living arrangements. It is high population, not density, that generates employment and other social opportunities that promote high levels of human activity.[3] Thus it is also a high population that puts people at risk of becoming infected, not density.

However, to say that density is completely unrelated to COVID-19 infection rate is also incorrect. The lack of correlation between density and the spread of the epidemic can be attributed to the fact that the different effects of density cancel each other out. Firstly, density increases the number of contacts between people and the likelihood of transmission of the disease in crowded places (which are a common occurrence, but not an absolute characteristic of densely populated regions).[3] Secondly, people in compact areas often have better access to online services such as grocery shopping and home delivery systems.[3] The people living in densely populated regions are also more likely to obey social distancing and other health advice than those living in suburban and rural areas.[3] As well, dense populations often have many more healthcare facilities and better healthcare infrastructure to service the large concentration of people.[3] All of these factors related to dense populations add up to cancel each other out, yielding an overall lack of relationship between density and infection rates.

From the Homeowner's Perspective

A large number of people are planning a move to the suburbs, as well many businesses are already taking the step towards moving their offices to suburban areas.[5] Due to this predicted pattern of future migration, urban planners may be forced to readjust to accommodate changing demands.

And what exactly are these changing demands? Experts predict that a major swivel in preferences lies in the type of housing that people will opt for in the future. As the epidemic continues and the importance of having personal living spaces, away from other people, the general shift in attitude has shown a preference for houses over apartments.[5] Suburban areas will likely experience a rise in demand for its housing, and urbanization is predicted to decrease as people move away from bigger cities.[7] During the epidemic, and most likely for a long while afterwards, multi-storey buildings will need to find ways to reduce contact with shared facilities such as elevators, doors, and other surfaces. As a result of health and safety requirements for future building design, high rise buildings would become more expensive to build.[7] Coupled with the decreased demand for apartments, it is likely that skyscrapers and high rise buildings will become less and less popular after the epidemic.[7]

As for the houses themselves, there will be less inclinations for open plan spaces - where the living room, dining space, and kitchen are unsegregated.[5] Instead, as people begin to see increasing separation between the 'outside environment' and personal living space, the entrance area will most likely be separated so that nothing is carried from the public areas into living quarters.[5] Water and air filtration systems, labeled as unimportant before the onset of the epidemic, will increase in popularity as people slowly realize their benefits.[5] Demand for such systems will likely increase dramatically as people move through the aftershock of the virus. This means that they will be more willing to pay for additional details to their home with independent water and air filtration systems

All of the future changes are linked to the fact that people's fear of infection will not go away after the virus does. The memory of sickness will prompt people to act in ways that provide them maximum safety and self assurance long after the sickness itself is gone. In this way, changes to building design and infrastructure are long lasting, and health-friendly design for homes will persist - eventually changing the way that private spaces are viewed permanently.

Implications for the Future of Public Spaces

In the wake of the 1954 cholera outbreak, The entire city of London, England was reconfigured to increase public health standards.[1] And if historical trends repeat itself in today's COVID-19 outbreak as it had done, then the world is due for massive changes to urban infrastructure in the near future. The increased conscientiousness about the design of public spaces would improve life long after the epidemic as urban design and architecture adopt the promoting health and sanitary practices as a major goal. Just as London's response to the cholera outbreak influenced the modern street grid design, so too will the changes that follow COVID-19 become long lasting adaptations that will benefit future generations to come.[1]

Firstly, designers will likely work more with pre-existing antibacterial materials such as fabrics and finishes such as copper. As well, industry experts now have adequate incentive to develop new materials that perform such functions, thus there is likely a shift in the materials of choice in the future.

Secondly, construction elements currently employed in healthcare facilities will likely be applied to other public spaces as design takes inspiration from what is already available.[4] For example, public spaces will mostly likely follow in healthcare facilities' footsteps in reducing the number of flat surfaces that germs can stay on, as well as installing ventilation systems that remove contaminated air.[4] Healthcare facilities themselves will likely get an upgrade as well. Hospitals during the epidemic have not been performing at their best in large part due to the building's incapability to accommodate a large number of sick people. Normally, this design flaw is on the backburner of society's agenda, however in the face of the epidemic, this flaw has become a fatal mistake, and society is more than ready to tackle the challenge. An example of future development for the design of hospitals is the focus on making normal hospital rooms more flexible in function, so that these normal rooms can be changed into ICUs and other useful spaces at a moment's notice.[4]

Before the epidemic, restrooms with doors in public spaces were already decreasing in popularity.[4] After the epidemic, this feature of public areas will most likely go entirely extinct. In addition to that, public spaces will likely shift towards adapting touchless technology such as automatic doors, voice activated elevators, hands-free switches.[4] These technologies were already available before the epidemic, however their marginalized status will likely change as mainstream demand shifted in a more health-conscious direction.

During the epidemic, online services have been gaining momentum and popularity as people find ways around quarantine regulations to continue their daily lives. The resulting boom in remote collaboration software has resulted in the facilitation of everything from remote schooling to working from home. With these advancements, there has been many speculations about the prospect of rethinking school and offices. Previously, shared workspaces were used more than remote work on the assumption that only physical closeness can foster better collaboration and higher work efficiency. However, if virtual working is successful, then these assumed benefits of shared working spaces will be proven incorrect. Potential decrease in open office working style will occur as different ways of collaboration are incorporated into the work environment. If a large enough audience begins to support online schooling and working from home, then it's likely that the future of school and office buildings is likely bleak as demands change. In addition, as more people find enjoyment in using online streaming and entertainment services, then demands for cinemas are also likely to decrease. All of these changes will sum up to change the demand for different types of buildings, resulting in a major paradigm shift in the world of architecture and design.

Biophilic Design

To the general public, the epidemic has revealed the importance of architecture in health promotion. Previously before COVID-19, not enough people considered making changes to public and private infrastructure when they want to live healthier. But now, the epidemic has brought to light the inextricable link between human health and the buildings we live in, the infrastructures we use. This sudden consciousness about immediate surroundings will no doubt transfer to other aspects of health promoting design. An example of this linked effect is biophilic design.[2]

Biophilic design is the incorporation of nature into man-made buildings. Whether it's rooftop gardens, or a green terrace, biophilic design has great effects in reducing anxiety in urban populations.[2] Engaging with nature in these small ways has profound effects on mental health, which in turn impacts the rest of the body to promote healthier immune systems.[2] This positive effect is being examined in children and the elderly, as researchers hope to discover new and different ways to decrease health risk.[2]

Due to the positive effect of biophilic design, as well as recent interest in tackling health issues by changing the immediate environment, it is likely that biophilic design will be a greater focus in the future of architecture and urban design.

There's already rising interest in biophilic design, the incorporation of nature into the built environment. Some people see this as a renaissance in design thinking and practice. Biophilic design removes or reduces anxiety in people, primarily through an emphasis on nature or design with natural features. In response to pandemics, researchers are studying syntactic relationships between children and nature, the elderly and nature.[2] Engaging with nature even just visually improves how we feel, affects mental health, so visual engagement may become more important.[2] Given that there could be a growing preference for proximity to nature, because we see it as more healthy and less of a health risk, it's likely that biophilic design will be of more interest to the design community.[2] It will probably become a greater part of the discourse in architecture, more mainstream, part of collective architectural thinking.

Conclusion

In sum, the epidemic has done a lot in bringing mainstream focus on health issues related to the way that homes and public buildings are designed. This will likely lead to a change in societal demand for different construction elements and materials used in future building design. As well, during the epidemic, online services have been adapting to provide people with alternative ways to go about their daily business. As people realize the efficiency of having work and schooling, as well as entertainment right in the comfort of their homes, it is likely that the type of public buildings in demand will also change. Places like office buildings and cinemas are more likely to go out of fashion as health concerns mix into other requirements of the general public. The effect of having increased health conscientiousness in architecture also has long lasting implications for the future. Experts predict an increase in demand for biophilic design, as well as instrumental changes to public infrastructure and health facilities that will benefit future generations to come.

Chapter 9

Negative Impacts of Covid-19

With the start of quarantine around the world much of the world's population moved indoors, staying in their homes. This caused the air to clear up as CO2 emission caused by factories and transportation decreased, beaches and bodies of water started to clear up and animals started to return to their habitats and roam around freely. Many people were optimistic that we were heading in a positive direction in terms of the environment and climate change. But, as quarantine rules eased and governments tried to boost up their economies once again, emissions started to go up again and we began to pollute once again. Many experts have noticed that even with the promising beginning our numbers have returned to pre-covid and now instead, we have added more problems as a result. Covid will lead the world to a future with more traffic, more pollution, changing government priorities and climate change that is now worsening faster. "We still have the same cars, the same roads, the same industries, same houses. So as soon as the restrictions are released, we go right back to where we were," says Corinne Le Quéré, professor of climate change at the University of East Anglia in Britain. Back during the 2007-2008 financial crisis, emissions dropped and then rose back up, many experts are worried that this will happen again[3]. An example of this can already be seen; China was the first country to reduce quarantine rules and the air quality improvements that were seen in February and March as manufacturing and transportation largely came to a halt have now disappeared. With the constant changes and the pandemic still on the rise in some countries it is too soon to have definite answers but there are many red flags we must pay attention to.

Changing Government Priorities and Easing of Laws

As countries begin to ease up quarantine rules and open up once again we are starting to see many changes. Many countries were forced to close up businesses and factories due to the health of their citizens and were hit with a decline in their economy, and now they are desperate to boost it back up. For example, a new coal project has started in northern China, this in the future will lead to multiple health and climate problems because these infrastructures are used for many years.[2] As new projects are getting approved and factories start to open again, pollution levels have returned to pre-coronavirus levels and in some places have surpassed. Similar actions can be seen happening around the world. Another example is in the U.S; industries such as automobiles, fossil fuels and airlines have been pleading for cash, regulatory rollbacks and other special requests which the government is complying to.[2] Another industry benefiting is the oil and gas industry in the U.S. The aid they have received includes tax changes, breaks on the royalties they have to pay to drill or mine on public land, as well as, access to the Federal Reserve's $6000-billion Main Street Lending program[2]. Lukas Ross who is a senior policy analyst for Friends of the Earth, an advocacy group says that "the program has already been modified specifically along the lines the oil and gas industry has requested." The Trump administration has also pushed many regulatory rollbacks which includes; suspending enforcement of air and water pollution regulations, suspended a requirement for environmental review, cut back the states' ability to block energy projects and public input on new mines, pipelines, highways as well as other projects[2]. With the newfound desperation to provide an economic boost many governments around the world have put environmental issues on the back burner and shifted their focus.

Waste Management and Recycling

Waste and recycling has greatly been impacted due to COVID-19, there has been an increase in waste production and a decrease in waste and recycling management. This reduction continues to pollute and contaminate our water ways, the air and our land. There are some cities in the U.S which have ceased recycling programs due to concerns related to spreading of the virus in these recycling centers.[5] In many European countries, such as Italy, have stopped sustainable waste management measures.[5] An example of this would be infected residents not sorting their waste. Single use plastics have once again increased, many companies have switched from advising customers to bring their own reusable bags to using single-use bags, others such as Starbucks, have put a temporary ban on the use of reusable cups.[5] Single use plastics can still house viruses and bacteria. With the lockdown many people started to online shop and order food more frequently, this has resulted in an increase of domestic, organic and inorganic waste. The increase in waste can lead to a variety of environmental issues, these included, but are not limited to; soil erosion, deforestation, and air and water pollution. Another source is the increase in medical waste, hospitals in Wuhan, the origin of the coronavirus, produce an average of 240 metric tons of medical waste per day during the outbreak, compared to a pre-corona average of less than 50 tons.[5] All around the world there has been an increase of PPE waste, this includes things such as gloves and masks. While there is no evidence of the survival of the SARS-CoV2 virus in drinking and wastewater China has ordered wastewater treatment plants to increase the use of chlorine as a means to strengthen their disinfection routines.[5] Increasing the use of chlorine in water can be harmful to people's health. A decrease in waste management and recycling can have an impact on the environment in a variety of different ways, these will be long-term changes.

Traffic

With the increase in social distancing, the need to stay 2m (6 ft) away from others and the limit to the number of people who can hang out together has made it hard for many to use public transportation, many people fear contracting the virus. This has led to people using cars instead of more eco-friendly options of transportation such as public transportation and carpooling with others. A transportation news website fears for a post-quarantine "carpocalypse". In China, the first country to lift the quarantine rules, has reached pre-coronavirus traffic[3]. Many people still have yet to resume commuting and travelling. In the past many cities and countries around the world have pushed to expand bike lanes to try to shift away from bus, subway and train use but now it is unsure if it will be enough. It is still early to have qualitative data to analyze but many experts believe that traffic will increase and this will create a lot more pollution.

The Amazon Rainforest

The Amazon forest has been used by the native people for food and resources for thousands of years. These resources included rubber, palm fruits, nuts and many different medicines[2]. Since the 1970's and 80's deforestation has begun. The deforestation is due to highways such as the Trans-Amazonas and the soy highway[2]. Illegal logging has continued and accelerated during Covid-19. According to satellite data from the space research agency INPE, since April 2020, 64 percent more land in April 2020 was cleared than in April 2019, this is true even with 2019 being the biggest year for deforestation in the Amazon in more than a decade. Brazil's president, Jair Bolsonaro has been an advocate for increasing commercial exploitation of the Amazon Rainforest[2]. There has been an increase in illegal logging, mining and cattle ranchers, all of them have faced little law enforcement and legal action. Ane Alencar, director of science at IPAM Amazônia, a scientific non-profit says

"You can do whatever you want in the Amazon and you won't be punished." The pandemic has been used "as a smokescreen, a distraction" to cover up the deforestation and to allow the destruction to continue. Brazil, the Amazon and its indigenous population has been one of the worst hit parts of the world. The deforestation has added a second crisis and the two issues are threatening to merge. Once vegetation has been cleared it is typically set on fire. Starting in July, once the vegetation has dried the thick smoke that it creates can lead to heart and lung problems to arise in those that live nearby[2]. This occurs every year but this year it is much worse with Covid-19, which is a respiratory illness. Additionally, those who are suffering from the illness can experience aggravated symptoms due to the smoke, this puts an additional pressure on the hospitals which are already struggling with the pandemic. The Amazon rainforest is an important resource to all around the world and is home to 10% of the world's species.[1] If the deforestation continues to occur it will decrease biodiversity, a loss of medical advancement and treatment - 90% of human diseases are treatable with drugs which include ingredients from the Amazon, rainfall will decrease creating a ripple effect, drastically changing the world climate along with many other issues that will arise.[1]

Changes to Make

It is still too soon to have definitive answers to all of the impacts Covid-19 has had on the environment but that does not mean we have to wait to make changes. It is important for everyone from the government to businesses to individual citizens to make changes in their lifestyle before it is too late. Certain questions arise for the government; what will their priorities be, will governments try to boost the economy by using old, polluting industries such as fossil fuels and oil or will they encourage a "green stimulus" and use funds to create jobs and fund businesses in areas such as clean energy and energy efficiency? It is important for them to use recovery funds towards a future which is low-carbon. When quarantine bans lift the global economy is going to rise once again and activity is going to return in most countries, and a short-term decrease in greenhouse gas emission is not sustainable. There are changes that individuals can make as well, for example if possible try to grow your own vegetables to limit the number of plastic bags used and online grocery shopping. Other changes that can be made are; using public transit with proper protection and social distancing, using reusable masks and other forms of PPE, if you do wish to shop online choose to shop ethical and environmentally sustainable, and making sure to continue to property separate your waste. There are many alternatives that can be made and many of them are small lifestyle changes, such as using reusable straws. We have seen such changes in the past but it is now important to continue these efforts once the lockdown rules are lifted and to expand our sustainable ways to better the environment and our future.

Conclusion

The arrival of the Coronavirus has forced many people to change their lives and it will continue to have an effect on the lifestyle of many. The virus has brought new issues for the environment as well. While there have been some positive effects on the environment such as clean beaches, reduction in greenhouse gases over countries such as China, the clearing of the Venice waters these are only short term and are not sustainable. With the 2008 financial crash we saw a decrease of 1.3% in emissions but as the economy recovered the emissions rose as well, experts predict something similar will happen if we do not choose to change our ways[3]. Once quarantine and lockdown rules start to ease up, life will return to pre-pandemic and the environment will continue to be endangered. While it is important to give our economies a boost there are many ways to do this, it can be done following the old, unsustainable ways or we can shift towards a more sustainable and eco-friendly future. As we continue to monitor and analyze the indirect impacts the pandemic has had on our environment we continue to learn about the increasingly negative impact that the pandemic has had on our environment.

Chapter 10

Environmental Impacts of Masks and Water Use

With the start of COVID-19 at the end of 2019 many countries advised and set lockdown regulations. As well there was an increase of personal protective equipment (PPE) with now not just medical professionals but everyone being required to wear them. With the easing of lockdown rules and regulations wearing face masks and gloves have become the new norm and this new normal will continue on for what some experts believe to be the next two or three years even if there is a vaccination for the virus. The most common masks used are N95, surgical and non-surgical masks and cotton masks. As well with the lockdown many people have increased their water use, cooking more, taking more showers and washing their hands much more often, for longer. Factories that produce these masks and PPE have also increased and the production and usage of these masks has had a negative impact on the environment. Many of these PPE's have not been disposed of properly even if they are recyclable and many cities, such as in the U.S have temporarily stopped recycling facilities in fear of spreading the virus to the workers. Increased water usage has raised concerns since our water is such a finite resource. While it is too early to have answers on the total impact the pandemic has had on the environment the early warning signs are there. As well, many scientists have been unable to do research as they normally would due to the regulations placed by the government. Regardless of this, it is clear that we have drastically increased waste production and this waste continues to pollute our land and our water and is dangerous for the health of our animals, plants and humans.

Increased Mask Usage

With the start of the pandemic many people bought and stored a variety of items such as toilet paper, latex gloves and masks. This panic buying led to shortages of a variety of items around the world, including masks. This led to a problem in the medical community where these masks were needed the most, and the shortage led to an increase in mask production. Companies such as Honeywell ramped up production of N95 masks in the U.S.[13] The demand required government planning and incentives to convert existing assembly lines to transition to making masks, this turned out to be very costly for every country to produce their own masks rather than support international trade. Production of masks requires the assembly of many different parts and various types of inputs, it is a complicated process. Masks are commonly made from polypropylene (the density is 20-25 grams per square metre) which is a polymer derived from petroleum oil and for the packing of the masks, paper pulp (forestry) is needed for cardboard.[14] Other common materials used in mask production are polystyrene, polycarbonate, polyethylene and polyester in surgical masks.[14] This has led to an incredible increase, for example; Chinese production increased six-fold at the end of February, meaning that about 116 million masks were being produced every day, and around 200 million per day by the end of March, meaning a ten-fold increase.[14] Many factories are running at 110 percent capacity. Since masks are all plastic based they have a long afterlife after they are thrown out, as well they are all made from liquid resistant materials. These masks end up in our oceans or in landfills. Another concern is the sheer amount of masks that are being used. Surgical masks are only supposed to be worn for one day and then disposed of, the use of single use masks in addition to hand sanitizer bottles and soiled tissue paper has led to a drastic increase in clinical waste in our environment. Wuhan on February 24th 2020 generated 200 tons of clinical trash which is four times the amount the city can incinerate per day and in the UK between the end of February to mid-April more than a billion items of PPE were given out.[12] The carbon footprint of these masks can be calculated. The footprint of one N95 mask is around 50 grams of CO2 equivalent, a cotton mask is around 60 grams of CO2 equivalent, in comparison a banana is around 80 grams.[8] The difference comes in how many times each mask is worn, the cotton mask can be worn multiple times while an N95 mask can only be worn once, after 30 days (not including washing the cotton mask) the CO2 equivalent produced by N95's is 1.8 kg and 25 grams for cotton ones.[8] As an example Germany is using around 17 million FFP masks per month which is 850 tons of CO2 for N95 masks generated per month.[8] To compare that much CO2 would be equal to 370,000 medium-sized steaks or driving 42,5000,000 km at the speed of 100 km/hr with a 8I car, that's circumventing the globe 1060 times.[8] This is theoretical data but it shows how the small changes that each person makes in their lives can have big impacts collectively. Impacts are already being seen even though we are far away from the end of the pandemic. Masks are ending up in remote locations such as the Soko Islands - a small cluster of islands south-west of Lantau Island.[12] OceansAsia which is an environmental NGO based in Hong Kong noticed that on the 100 m stretch of the beach there were piles of disposable single use masks. OceansAsia is a part of WWF's Blue Ocean Initiative and Gary Stokes, the founder and director, along with his team has been monitoring the ocean surface for trash, over the past years they would few the odd mask but now they are showing up much more often with new deposits coming in with each current. Stokes says, "Due to the current COVID-19 outbreak, the general population have all taken the precaution of wearing surgical masks. When you suddenly have a population of 7 million people wearing one to two masks per day the amount of trash generated is going to be substantial." The long lasting effects this will have are still unclear but it is sure that they will be left in the natural habitat of many animals, both land and water, this leads to animals choking, entanglement or ingestion and death on the plastic after mistaking it for food. Additionally to all the waste that has been built up we are putting frontline workers keeping the cities clean at risk since droplets of the virus can linger on the mask or other infections such as meningitis, Hepatitis B and Hepatitis C can develop on the masks.[12] WHO's health guidelines advises to throw out masks and soiled tissue into lidded litter bins and medical gear should be sterilised and burnt in incinerators. The burning of garbage leads to its own environmental issues, as well not all regions have the capacity to get rid of and burn the amounts of clinical waste that are being produced. Laurent Lombard from Opération Mer Propre, a French clean-up charity says, "There risks being more masks than jellyfish [in our ocean]." The masks are adding to an already big problem - about 8 million tonnes of plastics enter our oceans each year, in total there is about 150 million tons of garbage in our marine environments.[3] Some experts have calculated estimates, if each person in the UK alone is using a single-use fask mask every day for one year that creates 66,000 tonnes of waste and 57,000 tonnes of packaging.[3] It is clear to everyone that the amount of clinical waste produced had added an additional aspect to a pre-existing problem.

Future Steps to Take

Many maritime nations depend on their ocean economies which are dependent on the health of the ocean. In order to support the health of the ocean there are a few changes that individuals can make in their daily lives. Zac Goldsmith, Minister from the Department for Environment, Food and Rural Affairs in the UK recently said "Efforts to tackle plastic pollution can help us improve ocean health, tackle climate change, support biodiversity and build sustainable livelihoods," in a recent World Economic Forum webinar. The rise of the pandemic has diverted the attention of governments away from environmental issues and has pushed important meetings, such as the UN's COP26 climate change conference which was supposed to be held in November 2020 being postponed[3]. Individuals can opt to use cotton and reusable masks instead of single use masks when possible to help reduce the amount of waste which they produce and wash their hands more frequently overusing latex gloves. As well as choosing to shop locally instead of international can help reduce the waste which is produced through packaging. A chemist from the Italian government agency for new technologies, energy and sustainable economic development suggests that "Countries should try to develop products made of the same polymer, that we can trace and collect in sealed disposable bins, where they can be disinfected and recycled[11]." Additionally, the disposable PPE can break down and is a new source of microplastic fibers which animals ingest and then humans or other larger animals might ingest as well. This raises a concern for global food safety[4]. Another action that can be taken is by putting strict measures to curb the unyielding proliferation of plastic waste. The Marine Waste Project of the National Oceanic and Atmospheric Administration recently approved the Marine Waste Action Acts by the European Commission and is expected to start promotion and increase awareness campaigns towards particle pollution[4]. The use of single-use plastics is damaging not only for our environment but also for global health.

Increased Water Usage

With more people being locked in their houses the water consumption has greatly increased. In Saskatchewan officials have said that they have seen an increase of three percent in residential water use compared to this time last year. Since commercial and industrial use is not at full capacity they are using less, there has been a 40 percent decrease when compared to 2019.[10] It is expected that as things are going back to pre-pandemic the usage will rise again and go back to normal. In America, residential water usage has gone up 21 percent from the beginning of February compared to the end of April.[1] That means that on average 729 more gallons of water was used in April than in February (24.3 gallons per day) this study was done by Phyn, a water-monitoring company.[10] This rise was expected since people spent more time at home. To break down the 24.3 gallons, there was 21 percent more sink use, 20 percent more toilet use, 16 percent more shower use and 3 percent more washing-machine use.[10] In all of the states in America, New York had the highest average daily water consumption at 30.9 more gallons (28% more).[10] In April it was recorded that on average Americans visited their sink nearly 50 times a day, 10 times more than in February.[10] A representative from Phyn said, "It's logical, increased home water usage is a by-product of the COVID-19 lockdowns, and highlights the impact this virus has on the consumption of a precious and finite resource." It is definitely clear to see that average water use has gone up and will return to normal levels once regulations lift if changes are not made.

Future Steps to Take

Simple steps individuals can take is to try to save water in any way that they can. Even though individuals might want to run their washing machine more, take longer showers, keep the hose in your garden running for longer it is important to keep track of how much water you are using and to only use as much that is needed. According to WHO if we do not change our use, approximately half of the world's population will be living in water-stressed areas by 2025.[7] It is important to keep track of the chemical cleaning products that you are using, especially with the pandemic, as they will enter our wastewater system. For example try to use products that contain soaps, bleaches or alcohol when trying to kill the novel coronavirus rather than products that contain specialized disinfectants or nanosilver. When cleaning your car stick to the car wash because when you clean your car at home the dirty water containing salts, motor oil and other materials collects in our storm drains and will most likely enter into our rivers untreated. The key to being mindful here is to limit water use, choose the right disinfectant products and dispose of them properly.

Conclusion

The science is clear and unequivocal, if we do not change our ways we will be facing major environmental crises' in the future. With the pandemic our waste production and residential water use has increased, both waste production and water use are an important factor to consider when thinking about being a more sustainable and eco-friendly citizen. Small changes can be made to everyday life that will help to limit waste produced from packaging and PPEs as well as limit the waste of our freshwater resource.

Chapter 11

Lockdown and the Impact of Lower Overcrowding

The lockdown has forced people to shift indoors, it has put a halt on factories and businesses and travel. These new enforcements have created a ripple effect and have led to some positive environmental impacts. The pandemic has allowed for short term changes but scientists warn us that these changes are not sustainable for the long term. Before the pandemic many scientists argued that a decrease in economic activity could help decrease global warming and pollution which would then allow for our environment to heal itself. In previous pandemics such as in 14th century Eurasia and in the 16th-17th century North and South America these effects were already seen. Many such as the International Energy Agency (IEA) warn that the economic turmoil might even stop companies from investing in green energy.[6] Additionally, due to the lockdown and travel bans there has been a halt to environmental policy making and climate diplomacy, the 2020 UN Climate Change Conference being postponed to 2021.[6] An executive director from IEA, Faith Birol states " "the next three years will determine the course of the next 30 years and beyond" and that "if we do not [take action] we will surely see a rebound in emissions. If emissions rebound, it is very difficult to see how they will be brought down in future. This is why we are urging governments to have sustainable recovery packages." What the pandemic has done though is that it has created an opportunity to judge the anthropogenic intervention on qualitative degradation of environmental components at various scales. We have seen changes to our air quality through a deduction in greenhouse gas emissions and fewer flights, improvement to land quality where noise pollution has decreased and animals start to return to their natural habitats, water quality improvement, and an impact from food production changes.

Air

While there have been temporary improvements to air quality in many countries it is important to remember that as life returns to pre-pandemic levels we are already starting to see levels rise up once again and in some places are worse than before. Air travel accounts for a large sum of air pollution and now we have seen a decline in the two. In China, according to the Centre for Research and Clean Air, the first country to initiate lockdown, there was a 25 percent reduction in carbon emissions, that is 200 million fewer metric tons compared to the same period in 2019, and a 50 percent reduction in nitrogen oxide emissions.[6] Air pollution not only affects our environment but also the humans who have to breathe in that air, the reduction in emissions seen in China has saved an estimated 77,000 lives over two months according to one Earth systems scientist.[6] The declining emissions can be seen through a reduction in air traffic, oil refining and coal consumption. These reductions are due to economic downturns and should not be seen as a good thing according to Sarah Ladislaw, a member of the Center for Strategic and International Studies. This is due to the fact that China has already begun attempting to return to pre-pandemic rates of growth amidst trade wars.[6] Environmental impact will worsen as supply chain disruptions in the energy market will worsen. After lockdown scientists from the European Space Agency noticed a decline in nitrous oxide caused by cars, power plants and factories in the Po Valley, in northern Italy between the beginning of January and mid-March. NASA has been monitoring pollution levels in cities like Wuhan, China and has seen a reduction by 25-40%.[6] NASA used an OMI - ozone monitoring instrument - to analyze the ozone layer, NO2, aerosols and other pollutants. Even with the drop in NO2 levels the air quality in China did not meet the standard which is acceptable by health authorities, this is in part due to other pollutants such as aerosols still being released.[6] A study that was published in May 2020 stated that in April 2020 global carbon emissions were down by 17 percent and could lead to annual carbon emissions to go down by 7 percent, the lowest it has been since WWII.[6] The decreases in emissions mainly come from reduced transportation usage and a decline in industrial activity. Even with this decrease it was recorded that in May 2020 carbon dioxide concentrations were the highest it has ever been in human history.[6] The only change that might be long term and sustainable is the shift to virtual conferencing technology and widespread telecommuting. Constantine Samaras, an energy and climate expert states that "a pandemic is the worst possible way to reduce emissions, technological, behavioural, and structural change is the best and only way to reduce emissions." It was added by Zhu Liu, a student at Tsinghua University, "only when we would reduce our emissions even more than this for longer would we be able to see the decline in concentrations in the atmosphere." While there has been a decline in the fossil fuel industry due to the lockdown it is stated that more than half a trillion dollars worldwide is intended to be put into these high-carbon industries, according to Bloomberg New Energy Finance.[6] This is supported by preliminary disclosures from the Bank of England's Covid Corporate Financing Facility that billions of pounds from taxpayer money is intended to be funneled to fossil fuel companies, the amount according to Reclaim Finance the European Central Bank is approximately €220bn ($258.9 bn USD).[6] While there was a decrease in emissions and an improvement in air quality for a short while due to the lockdown these impacts are only short term. To have long term change we need to redirect where we get our energy from, an assessment from Ernst and Young showed that a stimulus program focusing on renewable energy can create more than 100,000 direct jobs across Australia and for every $1 million spent on renewable energy and exports, creates 4.8 full time jobs while fossil fuels only create 1.7 for the same amount.[6] According to Secretary-general, José Ángel Gurría, of the OECD club for rich countries, "seize this opportunity [of the coronavirus recovery] to reform subsidies and use public funds in a way that best benefits people and the planet."

Water

Covid-19 has led to some short-term positive impacts and some long-term negative impacts to our world's oceans and waterways. In Venice the water in the canals has cleared up and has experienced greater water flow. The clarity in the water came from the settling of the sediments in the water that previously were distributed by boat traffic.[6] Due to the pandemic there has been a decrease in demand for fish and the cost of fish has decreased, this means that fishing fleets are mainly sitting idle around the world. Rainer Froese, a German scientist has said that fish biomass will increase because of this and might lead to an increase in fish biomass.[6] In European waters, fish such as herring might double in their biomass. But, as of April 2020 signs for aquatic recovery are only anecdotal. As more people are spending most of their time inside more animals are being seen in areas they were not before. This includes jellyfish in Venice's canals and sea turtles laying eggs on beaches that were avoided before (the coast of the Bay of Bengal), this is due to less human interference and less pollution.[6] In the mining states of India with a lack of stone mining, the river quality in nearby areas has improved due to less dust being released[3]. An example would be the levels of total dissolved solid in river water beside crushing units decreased almost two times[3]. This shows that with the correct pollution management it is possible to restore our environment and ecosystems can recover at rapid rates. On the other, with an increase of clinical waste face masks, gloves and other PPE has ended up in our marine ecosystems, polluting the water and ending up on the shore of remote islands. This is harmful for animals because they might think of the plastic as food, ingest it and possibly choke or die due to it. The increase of plastic has also increased the amount of microplastic that floats around in our oceans, some experts think that this will lead to a global health crisis.

Land

Animals have started to come back to their natural habitats as human interactions decrease, at the same time the increased pollution has threatened their habitats. In the U.S, fatal vehicle collisions with animals decreased by 58 percent during the months of March and April.[6] In South Africa, lions can be seen sitting on the sides of roads that were normally used by safari-goers and in Yosemite national park, bears and coyotes have been seen around empty accommodations[4]. In Japan and Thailand, monkeys and deers have started to roam the streets as well because of a lack of tourists.[5] But, in Africa many people in some countries will see a surge in bushmeat poaching.[6] A member of the Nature Conservancy, Matt Brown stated, " When people don't have any other alternative for income [due to the lockdown], our prediction -- and we're seeing this in South Africa -- is that poaching will go up for high-value products like rhino horn and ivory." Myanmar has allowed for the breeding of endangered animals such as elephants, pangolins and tigers in June 2020.[6] Some experts believe that the deregulation of wildlife hunting might lead to a "new Covid-19." But, Gabon, a country in central Africa has decided to ban human consumption of bats and pangolins to try to decrease the spread of zoonotic diseases.

In America the Trump administration has suspended the enforcement of a few environmental laws via the Environmental Protection Agency, this has allowed polluters to ignore these laws if they prove that their violations were caused by the pandemic.[6] On the other hand, in India the lockdown has caused a decrease in pollution in mining states. For example the land surface temperature has decreased by 3-5 °C and noise levels are now below 65 dBA, in comparison they were more than 85 dBA pre-lockdown[3]. The decrease in temperature has shown scientists that industry induced energy footprint enhances temperatures greatly. Even though we have seen reductions in land pollution and animals start to return these impacts are only short term, as life goes back to normal our pollution and negative impact on ecosystems will continue if we do not change our ways.

Food Production

There have been changes to the food production due to the lockdown. For example, small scale farmers have started to use digital technologies as a way to directly sell produce and community supported agriculture and direct-sell delivery systems have increased.[6] This is leading towards a positive environmental impact as many smaller online grocery stores sell organic and locally grown food. Small scale farming, shopping locally rather than driving are both good for the environment. There has been a great increase in the number of online stores due to the pandemic and people being unable to leave their house to shop. Even though carbon emissions have dropped due to more online shopping, methane emissions have increased due to livestock. In comparison methane is a more potent greenhouse gas than carbon dioxide. And, in China it is not certain that the "wet markets" which are full of wild animals will be shut down, killing these animals is harmful to the environment and can be damaging to ecosystems[4].

Conclusion

The lockdown and a huge decrease in air travel has led to a decrease in overcrowding. Due to this, many short-term positive impacts can be seen but there are still negative impacts such as increased pollution in our oceans due to clinical waste. As well, many experts believe that emissions will return to normal once the lockdown is lifted. According to one expert, "The pandemic is fast, shining a spotlight on our ability or inability to respond to urgent threats. But like pandemics, climate change can be planned for in advance, if politicians pay attention to the warnings of scientists who are sounding the alarm." The pandemic has proved that our ecosystems can restore themselves and are somewhat resilient if not completely destroyed, this gives scientists some hope but the long-term impacts will depend on how governments and corporations respond to the economic crisis the pandemic has created.

Chapter 12

Increase in Automated Equipment Development and Usage

Due to Covid-19 many people have been forced to be in lockdown, many factories and businesses had to close their doors or stop production. As well, global trade has come to a halt as it becomes risky to ship goods with the increased risk of contracting the virus. Some areas that have been hit are the food industry, the mining industry and the clinical equipment industry. With these new challenges many are looking towards replacing humans with machines and automatic equipment. Many companies have shown interest in the use of Industry 4.0 technology, including artificial intelligence and 3D printing. It is not possible to predict the entire effect Covid-19 will have on these and many other industries but we can see some changes already happening. NYU Stern School of business professor, Arun Sundaranajan had to say, "Crisis can be sort of a catalyst or can speed up changes that are on the way — it can almost serve as an accelerant."

Increased Automation

One industry that has been affected is the mining industry. The increased use of technologies is enticing, this will reduce the number of workers and travel that is required. Ahmed, in a sector commentary says, ""The uptake of automated mine solutions including self-driving haul trucks and remote operations centers has been slow but steady." Change in the mining industry can be seen as far back as 2008 when a mining giant, Rio Tinto's Mine of the Future initiative started, shifted to more automated mine fleets.[2] About one-third of the mine haul truck fleet is automated and many of its Pilbara mining, ore handling, processing and logistics operations are done by automated machines and remotely supervised and operated from a control centre about 1200 km away in Perth, Australia[2]. A more recent example can be seen is Resolute Mining's Syama underground gold operation in Mali[2]. It is claimed to be the first fully automated mine, these mines operate with automated trucks, loaders and drills[2]. According to Resolute these mines can operate 24 hours a day, with a site operations centre overseeing it all[2].

The pandemic has shown the world the vulnerabilities of the global value chains (GVCs) and disrupted them. The GVCs are synonymous with globalization and are characterized by high interdependence between global lead firms and suppliers from all over the world.[5] Many countries have experienced shortages of important medical equipment such as masks, to combat the virus.[5] Nations and firms have experienced difficulties and are experiencing risks associated with protectionist national trade policies, such as high import tariffs causing equipment shortages from China in the United States because export restrictions have increased and caused supply shortages.[5] Another example is the bottleneck in the production of ventilators and the availability of certain components which are produced by global suppliers.[5] Many countries only produce certain parts rather than build the entire product and this results in high interdependencies. Companies such as Royal Philips rely on its wide network of global suppliers so it can continue its operations.[5]

For many years before Covid-19 countries have tried to push the start of automated equipment and replace human workers, the pandemic has just pushed this demand even more and given it a head start. To try to mitigate supply chain risks, increase flexibility and improve product standards many global lead firms have used Industry 4.0 technologies, even in developing countries such as Bangladesh, a major exporter of apparel, has started to replace humans with robots in response to increasing wages.[5] Some experts wonder if the pandemic will cause the reversal of globalization.

It is a possibility that Industry 4.0 technologies (examples include 3D printing and artificial intelligence) replace humans and fill in the gaps that are currently being seen in the market of PPE and ventilators. The U.S Food and Drug Administration can be seen as an example, it is working with the government and its public-private partners such as America Makes, and the National Additive Manufacturing Innovation Institute, to deal with the shortages in medical supplies, by using 3D printing to make ventilators valves and other parts as well as make face masks and plastics shields even though there are certain limitations to additive manufacturing.[5]

Some countries have used Industry 4.0 technology to test for Covid-19, the Republic of Korea is at the forefront for this, they have successfully limited the number of deaths linked to the virus.[5] One Korean company, Seegen which performs multiplex molecular diagnostics, used artificial intelligence based big data systems to develop a test for Covid-19 in only a few weeks, a process which usually takes a few months.[5] The test was then approved by the Korea Centers for Disease Control and Prevention within a week and the test was up and running.[5] The system uses automatic testing, the samples are analyzed by a machine rather than by humans, speeding up the process and reducing error risk and it is less likely to lead to contamination.[5] Korea has been used as a model to show others that it is possible to use robots over humans. Using technology over humans reduces the reliance on low-skilled, low-cost labour in manufacturing around the world. GVCs will start to become more regional, moving closer to key final consumer markets in China, the European Union, Japan and the U.S.[5] The length of GVCs will decrease as it will get rid of a few of the steps. Many companies, additionally, are expected to have rocky returns since the economy has suffered, an example of this is Chinese machinery production, the market was worth an estimated $512 billion in 2019 and is now expected to reduce by 10.3 percent.[6]

While many companies are looking to shift towards more automotive usage robotics production has had a mixed 2020. China has become the world's largest market for industrial robotics and the fastest growing market worldwide, for industrial robotics it is predicted that it will have a "sluggish recovery" in 2021 following a decline in 2020.[6] This is due to the fact that consumer confidence has decreased and the automotive sector is struggling to recover. The pandemic will also boost investments into the robotics and automotive industries, not just help the transition.[6] At the Brookings Institution they had to say, "Economic literature over the last decade shows that these investments are made especially during a crisis."

The retail industry is another industry that has been hit hard. An expert had to say, "If the epidemic intensifies, there's an opportunity there for companies that make retail technologies that automate what would otherwise be in-person human interaction." There has been an acceleration in the installation of kiosk ordering systems.[1] CEO and CTO of Attabotic, Scott Gravelle says, "Long term, the future got brighter for us; short term, it's a challenge." Some of the challenges they are facing include demonstrating their products virtually and generating venture capital conversations which have currently stopped.[1]

Another area to look at is the healthcare industry. Robots can be used for both precision and are an easier way to ensure that germs are not spread. For example, Aethon's TUG autonomous mobile robot is used to deliver medications, transport linens and meals and dispose of trash, after the pandemic demand for these robots has increased.[1] Sanitizing medical facilities is another area that automation can be helpful in. Companies such as Brain Corp, which were previously mainly designing sanitation robots for commercial spaces now have access to the medical industry.[1] According to Phil Duffy, vice president of product management and marketing at Brain Corp said, "[cleaning hospitals is] obviously a target market that could really benefit from robotics."

The pandemic has made a lot of people start to, or increase the amount of items they purchase online, and demands for robot delivery is increasing. Robots can deliver anything from groceries to medicine. FEMA, a government agency, along with many others are now interested in how drones can be used to make deliveries during the pandemic.[1] There are also other organizations such as hospitals which want to start using technology to deliver medicine, food, water and test kits to those who are in self isolation or are in quarantine.

Effects on the Environment

The increased push towards more machines and robots will have both positive and negative effects on the environment. The reduction of GVCs will help to keep things local, this can reduce the pollution that is caused by travel of goods and parts done by air, sea and train. As countries start to build local they require less from others around the world saving on the emissions they create. But, increasing machines can also create pollution. For one, once the machine breaks down and gets old it will become waste, not all of it can be reused and oftentimes it is not disposed properly. The amount of pollution it creates depends on the size of the machine and how its environmental impacts can be reduced.[3] It is also important to consider the pollution that is created when creating the machine/robot that is used. Extraction of minerals and materials creates a lot of pollution as well.[3] The annual growth of industrial waste is about 5-7 percent per year, this is estimated to increase with the increase of automotive use.[3] It is hard to judge the actual amount of waste that is produced annually since numbers are always based on estimates of volume rather than on material balance data. Once the machine is in use the environmental factors to consider are the impact of the magnetic fields around the machine, the noise pollution it makes, vibrations, loss of lubricants or coolants and run off of any sort, graphite powder and fumes which can worsen air pollution.[3]

Machines are going to be used more and more but it is important to think of sustainable design when designing the machines. Sustainable development means meeting the needs of the current generation without negatively impacting the future generations needs.[3] In order to design sustainably we must; minimize the usage of raw material, especially those which are non-renewable, to consider how to maximize efficiency while minimizing energy consumption, figure out ways to make part replacement easy, introduce monitoring and control of machine operations while making it easy to reuse parts and materials at the end of the machines life.[3] To work towards sustainable development some actions that can be taken include; using less non-renewable resources and increase the use of renewable resources, minimize pollution, use green energy over conventional energy sources, and increase environmental awareness in the public and with the workers.[3]

Conclusion

Plans to increase robotics and automotive use were in works before the pandemic, but, with the arrival of lockdowns and social distancing rules the use was accelerated as more and more companies demanded machines to replace human workers. The use of machines is a means to make more standardized products while risking the spread of the virus, this helps the countries economy, especially during a time where economies are dropping. As well, due to Covid-19 we have started to see the limitations of the GVCs, and they have started to shorten while countries have started to design and build locally rather than getting parts and materials from all over the world. Industry 4.0 technology such as 3D printing and artificial intelligence are becoming more common as many different industries such as retail and the medical industry have started to use them. It has been predicted that sales of robotics, while demand is high, will be low due to the loss many companies have experienced during the pandemic. The increase of robots adds another layer to the environmental impact humans are inflicting on the environment. Impacts from the machine start from the extraction of materials needed to build the parts all the way to how the machine is disposed of. There are a few changes that can be made but it will have to be a collaborative effort from all over the world to ensure sustainable design and sustainable development.

Chapter 13

Lifestyle Changes and Their Effect on the Environment

The new coronavirus, SARS-Cov2, started on December 12th, 2019 in Wuhan, China. The virus continued to spread and was announced as a pandemic by WHO on March 11th, 2020, on April 24th, 2020 the number of deaths worldwide was 195,313 and the number of confirmed cases was 2,783,512. It became clear to the world early on that the pandemic was going to change the world, it has had a massive impact on human health, people's lifestyles and the economy. One of the biggest changes was seen through social distancing, the use of PPE and the lockdown - isolation at home. Many people's everyday lives were changed, from their diet to their job. These lifestyle changes have not only had social and economic consequences but it has also had an impact on the environment and with many of these changes lasting into the future they will continue to have changes on the environment. There have been some positive and negative short-term changes and while it is too early to predict all the long-term changes the pandemic is going to have on the environment many experts predict it will negatively impact the environment.

Lifestyle Changes

There were many changes that came with the lockdown and social distancing rules. These changes included dietary habits and lifestyle changes. One study done in Italy measured the changes that occurred to daily life. Italy has been one of the hardest hit due to the pandemic, with 35,234 deaths and 203,326 confirmed cases (August 14th, 2020). The study took place between 5th April to the 24th of April and was aimed at investigating eating habits and lifestyle changes that occurred among the Italian population ages 12 and up.[4] It consisted of a questionnaire packed containing questions about demographic data, anthropometric data, dietary habits information (with a focus on whether the individual sticks to a Mediterranean diet), and lifestyle habits such as grocery, smoking, sleep and exercise. The study included 3533 participants, ages 12-86 years old with 76.1 percent of them female.[4] The data from the study showed us that: the perception of weight gain was 48.6 percent, 3.3 percent of smokers quit smoking, slight decrease in physical exercise in 38.3 percent of the individuals.[4] Those aged 18-30 adhered to a Mediterranean diet over those younger and older and 15 percent of participants started shopping from local farmers or organic to buy fruits and vegetables, this was particularly seen in the Northern and Central part of Italy where BMI was recorded to be lower[4]. This data can be extrapolated to other parts of the world, making rough estimates as many other countries, such as the U.S have also experienced a high number of cases and deaths and social distancing has been heavily enforced. From the survey done two major changes can be seen; one - staying at home (including online schooling, working from home and limitations of outdoors and in-gym physical activity) and the hoarding of food, due to panic buying and restrictions in grocery stores. Another change is the increased boredom and stress meaning an increased energy intake (eating more). This leads to overeating, particularly eating foods high in sugar or also known as "comfort foods".[4] The increased food intake has also led to an increase in food delivery.

In a survey done in USA individuals were asked about their lifestyle changes, the answers were collected and the results were; 77 percent of people stayed home more, 73 percent washed their hands more, 65 percent avoided public places like bars and restaurants, 56 percent travelled less, 52 percent shopped online more, 36 percent avoided public transport and 5 percent made no changes. All of these changes, such as shopping more online, not using public transport have an impact on the environment (more on this will be discussed in the section).[5] Many American have stated that the pandemic to them feels like the 2008 financial crisis, it is an event that will reshape society.[5] It has also shown many how shaky the structure that our society is built on and the things we take for granted such like the global supply chain, manufacturing infrastructure, and deliveries to supermarkets. Many people have turned towards social media and the internet to stay connected with their friends and families. Many have also started to use apps like Zoom or Microsoft teams much more often. People have started to become more resourceful with shortages of normal items such as toilet paper and this has also been shared on social media. Small changes can be seen due to this resourcefulness, such as rather than buying food from outside or a frozen meal from the grocery meal people have started to cook more at home, testing recipes in the new found free time they have.[2] Some people have even started to garden and grow their own vegetables and fruits in their backyards.[2] The pandemic has allowed people to utilize their free time to connect with past hobbies which were forgotten due to hectic lifestyles. A catalyst is the shift of many companies allowing their workers to work from home and the smaller shops/companies have been temporarily closed.[2] While lockdown rules are starting to be reduced, routine temperature checks and thermal imaging cameras are starting to show up more and more in buildings.[2] People have started to use their backyards more as well, Tosso, a maker of yard games stated that sales in April 2020 increased by 134 percent.[7] Activities that took place indoors or in public places can be promoted to be done in your backyard. Many people have decided to do home renovations as well, searches for flooring and doors rose by more than 50 percent and foot traffic in Home Depot rose 10 percent while at Lowe's it rose 21 percent in a period from February to April.[7] People have started buying a lot more to cure their boredom, sewing sales have gone up by 140 percent while board games have gone up 187 percent.[7]

Another lifestyle change is the use of public transportation. While in the main heat of the pandemic many public transportation methods such as buses and trains were required to reduce their capacity to as low as 15 percent. Even if the small number of commuters were forced to use a car instead, traffic congestion worsened. There are cities around the world that are trying to encourage people to walk and bike instead to work and some are reassigning road space to add more bike lanes and widen pavements. If more people started to walk or bike to work this would be better for the environment and we would be workers towards a greener commute.

At the start of lockdown it was announced that gyms would be closing and many people had to change the way in which they practice a healthy lifestyle. In the U.S sales of home gym equipment increased by 130 percent in March 2020, compared to last year.[7] Free weight sales increased by more than 180 percent and stationary bike sales from a company named Peloton recorded that a record breaking 23 000 people joined their online live stream fitness classes.[7] The investments in home gyms will offset gym memberships in the future as more people shift towards working out at home or in the environment (hikes, swimming, etc.). As well, many people have joined online fitness programs or classes.

One of the biggest changes to come from the pandemic is the increased hygiene consciousness, people have started to wash their hands more often, use hand sanitizer more often and use PPE (masks, gloves, face shields, etc.). A study done in India by the IANS C-VOTER Gallup International Association Corona group roughly 87.2 percent of Indian started to look after their personal hygiene much more[3]. As well, many brands for health and hygiene have started to increase their production and advertisements taking advantage of the pandemic.

In Canada many examples can be seen as well of the lifestyle changes. According to a survey done by the Association of Canadian Studies, 55 percent of Canadian participants bought at least one item online for the first time or started to purchase online more frequently.[1] This trend will force smaller local shops to create websites and increase their online presence. In Canada a lot of online learning can be seen as well, about 25 percent of Canadians have completed some sort of learning course or tutorial online according to the survey.[1] Christian Leger, executive vice president of Leger who helped conduct these surveys said, " For a lot of Canadians, e-learning is becoming part of their lifestyle." Almost 50 percent of the Canadians participants have been working from home, Baroque added, "Employers will need to react, employees will need to adapt because working from home is here to stay."[1] An additional 80 percent of participants said they were happy working from home.[1] According to Baroque many Canadians will shift towards home offices in the future, leading to office buildings down-sizing and even clothing stores will decrease with the shift to online.[1] How people spend their money has also changed. People have stopped spending on "non-essential" items, according to Baroque a new attitude, "cool to be cheap" has emerged.[1] In the survey 32 percent of Canadians reduced their spending on anything they consider non-essential.[1] It is advised to Canadian to consider their short term and long term saving strategies. In terms of food, many Canadians have started to go back to staple products and have started to cook more at home, something that was not seen before the pandemic. This will impact on what is stocked in grocery stores. How much people are spending has changed as well, during the first month of the lockdown many Canadians took part in panic buying and about 25 percent of survey participants followed pantry loading, buying in bulk.[1] The pandemic caused flights to be cancelled, about 50 percent of Canadians surveyed had to cancel or change vacation plans. "Staycations will probably become the norm for 2020," stated Baroque. This is very good for the local tourist industry. In total, for the majority of people the pandemic has turned their lives upside down and has caused a complete lifestyle change.

Impact on the Environment

There will be many short-term impacts of these lifestyle changes, but what people do once the lockdown is lifted will determine the long-term impacts. Many individuals have noticed environmental changes in their local environments such as; cleaner air, safer, calmer roads with increased wildlife roaming around. This has given many a glimpse into a greener world and what it might be like to live in one. Because of the many lifestyle changes such as working from home, travelling less and just staying in one place has decreased levels of carbon dioxide and nitrogen dioxide. Satellite data has shown a decrease in atmospheric levels of nitrogen dioxide over cities and industrial centers in many European and Asian countries, in some places falling by 30 to 40 percent compared to last year[2]. It has been estimated the slowing of the economy due to the pandemic will reduce global carbon dioxide emissions by 8 percent for 2020.[2] To show the bigger picture, if we wanted to reduce to less than 1.5C above pre-industrial temperatures, as was decided on in the Paris Agreement, globally we would have to reduce emissions by this amount every year[2]. On the other hand, an increase in online shopping for groceries and retail items and an increase in the use of PPE and personal hygiene products has increased the amount of waste that we produce. This waste has ended up in our oceans and landfills causing harm to the animals that live there and negatively impacted the ecosystems. In order to combat climate changes we need governments to make decisive, pro-active decisions, coordinated internationally and to ensure we keep levels of atmospheric pollutants down. We need to continue to make changes in the future and not just go back to the way life was pre-pandemic.

Conclusion

There have been many lifestyle changes, to list a few; decreased use in public transportation, increased online shopping, increased online schooling, more working from home, etc. All of these changes have caused major lifestyle changes to occur for citizens all over the world. Our actions greatly impact the environment. There have been some positive and some negative impacts from these changes, in order to maintain the positive changes in the long-term people will have to continue to work towards decreasing their waste and their pollutants.

Chapter 14

Differences in the Developing World and Developed Countries

Developing nations have consistently reported low rates of infection and deaths. However, this is not an indication of their success at combating COVID-19, rather these low figures are indicative of inadequate testing capacities and rampant unreported infection.[4] Developing countries are often plagued by unstable political, health, and economic infrastructures. Any one of these factors can be devastating in a normal year, however during the epidemic of 2020, these weaknesses have been especially fatal to the citizens of developing countries.

While it's true that developing countries generally have a younger population, and that younger people have been shown to have more resistance to the COVID-19 virus, it is still completely wrong for a developing country's leaders to assume that they are safer due to the age distribution of their countries.[2] This is because the studies on the effect of age on risk of infection are conducted in first world countries, on a population with access to adequate nutrition.[2] For the youths of developing nations, improper nutrition in the past and in the present could potentially affect their immune systems to increase the risk of catching and succumbing to the virus.[2] Yet despite this, some leaders of third world countries are actively banking on the fact that having a younger population will slow infection rates.[2] As a result, these leaders are greatly underestimating the risk of COVID-19 and are not taking proper preventive action against a worsening epidemic at their doorsteps.

As of now, there are three projected schemas for combating the virus: developing herd immunity, implementing vaccination, and using the testing-tracing-isolation method.[1] Herd immunity refers to the process of building up population immunity against the virus. It requires letting the infected and those without immunity to the virus die out, leaving a population composed mostly of people who are immune to the virus.[1] Needless to say, herd immunity is effective but only at the very high cost of millions of lives lost and utter social devastation that follows. The second method of dealing with the virus - vaccination - is implausible at the current moment since an effective vaccine is still yet to be developed.[1] Waiting around until the vaccine is developed and fully tested is not a viable option as people are still dying in the meantime. Even when a successful vaccine is available, developing nations simply lack the proper resources to both acquire and distribute enough vaccines to its people.[1] From this, it is easy to see that the most effective as well as the most plausible method of dealing with the virus is to test, trace, and isolate sources of infection from the rest of the healthy population. This is exactly the sort of action that countries such as China are advising the rest of the world to do.[1] However, one fatal weakness to this advice is that it assumes all countries have the testing capacities as well as the proper governance efficiency to carry out such complicated processes.[1] Unfortunately for the developing countries that were having problems with political stability and healthcare implementation even before the virus, this trace and isolate method is simply unfeasible.

That being said, the governments of developing nations are still trying to do their part in controlling the epidemic. They do so by following in the footsteps of the rest of the world in imposing state of emergency, prohibiting public gatherings, shutting down schools, closing down all businesses except the care essentials, and banning travel.[4] However, the way that the economy and the society of developing countries react to these procedures is far different from developed nations.

Consequence of Lockdown in Developing Nations

In developed countries, welfare systems and government subsidies act as safety nets during quarantine to prevent lockdown procedures from causing or even exacerbating poverty.[1] However, in poorer countries, these welfare systems are either non-existent, or faulty enough so that the people in lockdown are going hungry.[1] Additionally, developed countries are able to provide relief packages and other aid to businesses to act as economic stimulus that safeguards against bankruptcy.[1] Like the predicament with inadequate unemployment welfare, developing countries simply don't have the resources to help businesses stay afloat, as well as adequate governance systems to implement the help. As a result, many businesses in developing nations are taking a large hit during the pandemic, and more and more are sinking into the throes of bankruptcy as the quarantine continues.

Asides from lack of unemployment welfare, another reason why the citizens of developing nations are finding quarantine especially difficult is due to the industry distributions of developing nations. Third world countries specialize in mostly agriculture and production, both of which require on-site attendance of workers. Quarantine and social distancing measures greatly decrease the number of workers in these industries, and in doing so decrease the productive output of businesses working in these fields.[1] In addition, developed nations simply lack the technological infrastructures to allow effective communication when working from home.[1] This lack of capability to work from home prevents people from earning wages during quarantine in developing nations, thereby putting households at risk of poverty. In sum, citizens of developing nations have been finding it difficult to hold down jobs amid quarantine and social distancing measures due to a combination of factors. Everything from the type of industry they work into the country's undeveloped information infrastructure all play a role in exacerbating poverty as people lose their jobs amidst quarantine. On a larger scale, businesses in the agriculture and production-based industries of third world countries are also negatively impacted by quarantine. Therefore, it is by no means a far stretch to say that lockdowns in developing countries are increasing poverty, hunger and destitution.

These economic impacts are mostly felt by the younger, working populations of these developing nations. Young people - especially young women - lack access to savings, and work in economic sectors that are especially sensitive to distancing restrictions.[1] In the long term, the existing gender and social inequalities may be exacerbated by prolonged quarantine.[1] Combined with the loss in household income, quarantine may increase poverty and eradicate the progress third world countries are trying to make towards social and economic equality.

Other long-term effects of lockdown include its negative impact on human capital in developing countries. Because of the insufficient informational infrastructures of third world countries are incapable of supporting working and learning at home, students who are incapable of going to school during the epidemic are missing large portions of their school year.[1] As the quarantine continues to drag on, the cut off from schooling could lead to a massive decrease in the supply of skilled workers and professionals for the future, creating massive setbacks to the burgeoning pace of technological development of third world countries.

Inefficiency of Health Interventions in Developing Nations

The implementation of quarantine itself is problematic and sometimes even downright impossible as third world countries are finding it harder to reinforce regulations. This apparent weakness is especially visible in the slums and refugee camps in these third world countries, where there is barely enough space to fit all the people, much less to practice social distancing.

In areas that are able to provide healthcare coverage for communities, the healthcare system of third world countries has been buckling under the sheer number of COVID-19 cases.[2] The lack of space to accommodate patients, along with the proper equipment for treatment, has been wreaking havoc on the healthcare facilities of third world countries as they struggle to treat the sick without infecting healthcare workers.[2]

All of this is made worse by poverty and lack of resources in poorer neighbourhoods of developing nations. In countries where citizens have inadequate access to water, the people must face the epidemic with no handwashing and other basic health interventions.[2] Masks and PPEs are another issue altogether. Not only do third world countries lack the technology and the expert personnel to commit to their own large-scale production of PPEs, they also often don't have enough resources to buy healthcare equipment. The worldwide demand for masks is still climbing, and much of what is on the market is taken up by the countries that can afford it. The lack of protective equipment and masks puts the healthy people at risk, but most of all it threatens entire healthcare facilities as healthcare workers struggle to do their duties safely without the proper equipment to protect themselves from infection.

Collapse of Global Demand Amid the Epidemic

As countries all over the world closed borders amid the epidemic, world trade grinded to a halt. For the third world countries that depend heavily on the income from exports, this particular side effect of the COVID-19 virus is especially devastating.[1] Massive cancellations of orders from the world market has resulted in equally large-scale decrease in job opportunities as businesses struggle to cut their losses.[1] For example, cancelations of fashion retail orders on the world market has left the garment sector in shambles.[1] By March 2020, in Bangladesh alone, a quarter of the 4 million people employed in the garment industry had found themselves without jobs as exports from Bangladesh decreased by more than 80% from the year before.[1]

Aside from production, tourism is also an area of hot demand from third world countries, in particular small island developing states.[1] It goes without saying that travel restrictions are taking quite the toll on the tourism industry as demands plummeted during the epidemic.

Both tourism and the exportation of goods have helped developing countries by increasing their income as well as by integrating them into the greater world market.[1] Before the epidemic, third world countries have been making progress towards integrating themselves into the world stage, using the venues of trade and tourism as engines for growth.[1] Developing nations are so called 'developing' because they are on a trajectory towards graduating from their third world status. However, the sudden decrease in global demand amid the epidemic has become a major roadblock to this progress of growth, condemning developing nations to sink further into economic turmoil and poverty.[1]

As businesses close down or start to implement social distancing, the global demand for migrant workers from developing nations has also been decreasing.[1] These migrant workers, suddenly jobless amidst a global epidemic, often return to their home countries. This influx of home-bound migrant workers from abroad can put further pressure on already buckling healthcare and social protection systems, thereby exacerbating the economic and social issues faced by developing nations during the epidemic.[1]

A Note on the Response of First World Countries

First world countries have developed healthcare systems as well as adequate resources to combat the virus and implement preventive measures. Thus, it was expected that first world countries would struggle with but still eventually succeed at eradicating the virus. What the world has seen instead is climbing rates of both infection and deaths in first world countries as political leaders failed to apprehend the seriousness of the threat.[3] This failure of first world countries is entirely political. Entire countries ignored the threat of COVID-19, and even as the climbing infection rates forced the first world countries into action, political leaders sent mixed signals to citizens.[3] These unclear messages not only encouraged many people to act in ways that worsened the spread of infection, it also generated a lot of internal friction as people disagreed about everything from the seriousness of the threat to what actions to take.[3] This internal friction has become the main roadblock to the progress that first world countries are trying to make.[3] Thus, the first world countries were not only slow to react, they were also slow in their actions.

Prominent examples of this failure to act are the United States and Britain.[3] The hobbling of the affordable care system in the United States was mirrored by spending cuts to the National Health Service.[3] In addition, much of the pandemic prevention procedures and offices in the United States were either weakened or done away with altogether by the current acting president.[3] As a result, both countries failed to equip their hospitals and health workers with the needed equipment, causing losses that would've been entirely avoidable.

Chapter 15

Conclusion

It has been well established that COVID-19 and the environment have between them a complicated two-way link. The way that humanity has been reacting to the COVID-19 pandemic has profound effects on planet earth. As mentioned before, lockdown procedures that limit human activity will have a significant impact on aspects of the immediate environment such as air quality and biodiversity. These positive effects of lockdown will increase environmental sustainability and overall human health respectively, giving people great hope for the future after COVID-19. As humans recede into quarantine, more wild animals have been left with more breathing space, thereby increasing the biodiversity and the overall ecosystemic welfare of planet earth. In addition, the decrease in factories and cars that actively release air pollutants has decreased levels of air pollution. As the virus continued to spread to different countries around the world more of them started to implement lockdown and social distancing rules. The lockdown has led to some positive short-term impacts such as reduced carbon dioxide and nitrogen dioxide emissions over countries like China and the decrease in tourists and overcrowding have led to the canals in Venice to clear up along with other effects. The reduction in flights has also led to reduced emissions. On the other hand, some major climate conferences such as the 2020 UN Climate Change Conference has been postponed to 2021. There have been changes in food production as well, some people have started to shop more locally and organic as well as, have started to cook more at home, this is good for the environment as it reduces travel emissions. The next three years are crucial in determining what happens in the next few decades and if we do not take action our emissions will return to pre-pandemic levels. In general, animals have started to roam around more freely in places they previously did not go to. In order to keep the positive momentum going it's important for governments to enforce more environmentally friendly rules and law.

Despite the optimism presented on social media, with videos of river dolphins and other wildlife frequenting the spaces where humanity has retreated, the overall outlook for earth's biosphere in light of the pandemic is grim. There have been many negative impacts due to the pandemic. This is because we still have the same number of cars, the same roads and the same industries. Many experts predict that once the lockdown rules have been lifted life will resume as it was pre-pandemic. Many governments are trying to boost their countries economy, this has caused them to approve many new projects such as in China, a new coal project. In the U.S, many industries have been offered aid, like the Federal Reserve's $6000 - billion Main Street Lending Program. Many restrictions have been lifted, with companies using the pandemic as an excuse to their actions. Many countries have reduced waste management and recycling programs in an effort to reduce the spread of the virus in these facilities. With the reduction of public transportation many people have started to use their own cars, this has led to increased pollution and congestion on our roads. The pandemic has been used as a cover up to allow illegal logging in the Amazon Rainforest, this rainforest is a crucial ecosystem and if it is destroyed it will lead to great environmental consequences. In order to combat this it is important to work towards using more green energy and trying to produce less waste. And it must not be forgotten that lockdown is only temporary, people will soon emerge from quarantine to push wildlife back to the fringes, and air pollution levels will climb back to what it was before as humanity hurries to restart its businesses and economies. As well, there are other negative impacts of COVID-19 such as water availability, deforestation, and waste production that are proven to be detrimental to human health. These negative implications for human health are long lasting and far outweigh the transient positive effects, indicating a not so bright forecast for the future after COVID-19.

It is clear that people all over the world have started to use masks more often, there has been an increased demand for personal protective equipment (PPE), The demand has forced companies to work overtime to keep up, the increased factory use has increased the pollution they create. As well, the materials the masks are made out of are harmful for the environment. Many people have chosen to use single use masks over cotton and reusable ones which has added to the clinical waste in our oceans. This has a negative impact on our marine ecosystems, endangering the animals and plants that live in them. There also has been more water usage, people have started to wash their hands more often, take longer showers and wash their clothes/dishes more often. Water is a finite resource and it is important to use it responsibly. It is predicted that water usage will return to normal once the lockdown rules are lifted.

As well, the environment is also influencing the spread of the pandemic. The natural environment may play a role in the current COVID-19 epidemic via directly influencing rates of infection, while the man-made environment is impacting the population's response to the pandemic.

Experts have predicted that everything from exposure to UV radiation, to temperature, to humidity plays a role in the transmission of the COVID-19 virus. Contrary to popular belief, the virus will not go away in warm weather, even though it is proven to have a preferred temperature band for optimal virus survival. In this manner COVID-19 is similar to its predecessor SARS CoV-2 in that both have tendencies for seasonal patterns of fluctuation rather than showing up as an all-at-once global sweep of infection. As for humidity, both too-high levels as well as too-low levels of humidity have been shown to exacerbate the spread of the virus. This means that for Northern Hemisphere countries such as Canada, the move to the dry and cold winter season is likely to see an increase in infection rates. However, the impact of environmental factors on infection rates is much less than initially anticipated. Current and ongoing research indicate that the effect of environmental factors are nearly nonexistent when compared to the effect of health measures such as social distancing. This is mainly because most cases of infection were transferred indoors. These research findings support the current message about the absolute importance of obeying quarantine guidelines and other safety measures in attempting to stop the spread of COVID-19.

COVID-19 infection rates as well as other aspects of its impact has been drastically different in developing nations. This is in part due to the developing world's inadequate health infrastructures, which makes it hard to gather the sick and implement other healthcare interventions to prevent further spread of the virus. In addition to that, the water shortage in some developing countries means that their citizens are incapable of performing acts of basic hygiene such as washing hands. Another reason why the developing world is handling the epidemic badly is because of the third world country's incapability of dealing with quarantine. The lack of unemployment welfare coupled with information infrastructure that isn't developed enough to support working from home, more and more citizens of third world countries are faced with the risk of poverty as the quarantine continues. The insufficient information infrastructures also mean that students who are out of school during the quarantine cannot attend class from home. These massive breaks away from the learning environment will no doubt cause a decrease in human capital, thereby pushing third world countries back a step in their progression towards growth. Overall, bad living conditions during quarantine coupled with buckling healthcare facilities in third world countries are exacerbating the effect of COVID-19. On a final note, proper social distancing and self isolation is impossible to implement in the slums of third world countries. This effectively turns the slums into a massive brewing pot for the virus, further prompting infection rates and possibly even viral mutation. Thus, the different environment of third world countries may be a contributing factor to the seriousness of the COVID-19 pandemic within their borders.

Finally, it is no question that such a large-scale pandemic will change the way that people live, both in the present as well as in the future. COVID-19 has led to a lot of paradigm shifts as lifestyles are disrupted during quarantine. Overall, the epidemic has done a lot in bringing mainstream focus on health issues related to the way that homes and public buildings are designed. This will likely lead to a change in societal demand for different construction elements and materials used in future building design. As well, during the epidemic, online services have been adapting to provide people with alternative ways to go about their daily business. As people realize the efficiency of having work and schooling, as well as entertainment right in the comfort of their homes, it is likely that the type of public buildings in demand will also change. Places like office buildings and cinemas are more likely to go out of fashion as health concerns mix into other requirements of the general public. The effect of having increased health conscientiousness in architecture also has long lasting implications for the future. Experts predict an increase in demand for biophilic design, as well as instrumental changes to public infrastructure and health facilities that will benefit future generations to come.

As for the lifestyle of the common citizen dealing with the pandemic, the shift to quarantine has been difficult for many people, there have been many lifestyle changes and these changes have had an impact on the environment. Some of the lifestyle changes include, the limiting of public transportation use, the shift to working at home, dietary and fitness changes and personal hygiene changes. A study done in Italy showed that because of boredom many people have started to eat more and have reduced their fitness. Surveys done in America have shown that a lot more people are shopping online. Similar results were seen in Canada. Increased online shopping has led to a greater production of waste, as well many people buy from all over the world, the environmental cost of buying from overseas is huge. But, many people have become resourceful, starting gardens in their homes and cooking at home instead of ordering food which is good for the environment. Some people have started to pick up on hobbies lost in the past such as sewing and playing board games. Many of these lifestyle changes will continue into the future once the lockdown rules have been lifted.

Perhaps the most important effect of the pandemic on human activity is how the pandemic is changing industries. There has been an increase in robotics use and automotive equipment. As many workers were forced to work from home and as factories were shut down companies started to look for alternatives and replacements. Many have turned towards Industry 4.0 technology, this includes 3D printing and artificial intelligence. Before the pandemic the shift towards using more robots had already begun but Covid-19 has acted as a catalyst pushing it forward faster, One industry making major changes is the mining industry, companies such as Resolute Mining have started to shift their entire production to automated machines controlling them through a central control center. Another industry is the medical industry, making hospitals have looked into using robots in order to maintain sanitation and decrease the spread of the virus. These robots will perform tasks such as delivering medication and food to patients. The pandemic has shown the world how fragile the global value chains are. Production has been difficult with the limitations set by some countries on export and import. Since the creation of an item often involves parts of it coming from many different countries, items such as ventilators have been difficult to produce. The shortcomings of the GVC have pushed countries to design and build locally, this will reduce travel costs, keeping it local is good for the environment.

Throughout our analysis of the pandemic, the environment, and humanity's position between them, we have seen over and over again the interconnectivity between humanity and its environment. There is no doubt that the interactions between human and nature, as well as the inter-relations between human and human-made environments are complex and multi-faceted. However, through our analysis we hope to communicate the bi-directionality of those interactions, and raise awareness of the effect that humanity has on the world around us. We especially want to draw focus to the fact that the way humans shape the environment turns around to shape humanity. This human to environment to human chain of reactions may serve to perpetuate such crises like the COVID-19 pandemic. As we hope we have demonstrated, these chain reactions not only leave remarkable effects on life in the present, but also have long lasting effects far into the future. Through our attempt to raise awareness we hope to also inspire action. Issues such as water waste and medical waste disposal are challengeable issues, waiting to be recognized and solved. The only way to deal with the pandemic, and with its side effects is through collective effort on humanity's part. Instead of placing blame and pointing fingers, the world and its inhabitants should focus on uplifting the vulnerable populations, and enacting change for a better future.

References

Chapter 2

1. Atmosphere Monitoring Service. (2020, April 6). CAMS tracks a record-breaking Arctic ozone hole. Retrieved from https://atmosphere.copernicus.eu/cams-tracks-record-breaking-arctic-ozone-hole
2. Australian Government. (n.d). Montreal Protocol on Substances that Deplete the Ozone Layer. Retrieved from https://www.environment.gov.au/protection/ozone/montreal-protocol
3. Le Quéré, C., Jackson, R.B., Jones, M.W. et al. Temporary reduction in daily global CO2 emissions during the COVID-19 forced confinement. Nat. Clim. Chang. 10, 647–653 (2020). https://doi.org/10.1038/s41558-020-0797-x
4. Rai, S. (2020, March 30). The Ozone layer is healing but it's not because of Covid-19 lockdown. Retrieved from https://www.esquireme.com/content/44936-the-ozone-layer-is-healing-but-its-not-because-of-covid-19-lockdown
5. Rose, K. (April 13, 2018). Does CO2 Deplete the Ozone Layer? Retrieved from https://sciencing.com/co2-deplete-ozone-layer-4828.html
6. University of East Anglia. (2020, May 19). COVID-19 crisis causes a 17 percent drop in global carbon emissions. ScienceDaily. Retrieved from www.sciencedaily.com/releases/2020/05/200519114233.html
7. Wuebbles, D. (2020, May 26). Ozone layer atmospheric science. Retrieved from https://www.britannica.com/science/ozone-layer

Chapter 3

1. CREA. (2020, April 24). How air pollution worsens the COVID-19 pandemic. Retrieved from https://energyandcleanair.org/publications/how-air-pollution-worsens-the-covid-19-pandemic/
2. Hoang, U. (2020, May 20). Is there an association between exposure to air pollution and severity of COVID-19 infection? Retrieved from https://www.cebm.net/covid-19/is-there-an-association-between-exposure-to-air-pollution-and-severity-of-covid-19-infection/
3. Isphording, I., & Pestel, N. (2020, June). Pandemic Meets Pollution: Poor Air Quality Increases Deaths by COVID-19. Retrieved from https://www.iza.org/publications/dp/13418/pandemic-meets-pollution-poor-air-quality-increases-deaths-by-covid-19

4. NOAA. (2016, April 13). What are microplastics? Retrieved from https://oceanservice.noaa.gov/facts/microplastics.html
5. Singhal, S., & Matto, M. (2020, April 29). COVID-19 lockdown: A ventilator for rivers. Retrieved from https://www.downtoearth.org.in/blog/covid-19-lockdown-a-ventilator-for-rivers-7071
6. Winter, L. (2020, July 14). Analysis Links Poor Air Quality to Increased COVID-19 Deaths. Retrieved from https://www.the-scientist.com/news-opinion/analysis-links-poor-air-quality-to-increased-covid-19-deaths-67738
7. Yunus, A., Masago, Y., & Hijiokaba, Y. (2020, April 27). COVID-19 and surface water quality: Improved lake water quality during the lockdown. Retrieved August from https://www.researchgate.net/publication/340956089_COVID-19_and_surface_water_quality_Improved_lake_water_quality_during_the_lockdown

Chapter 4

1. Butler, R. (2020, April 11). Despite COVID, Amazon deforestation races higher. Retrieved from https://news.mongabay.com/2020/04/despite-covid-amazon-deforestation-races-higher/
2. CDC. (2020, June 22). Animals & COVID-19. Retrieved from https://www.cdc.gov/coronavirus/2019-ncov/daily-life-coping/animals.html
3. Dupont, K. (2020, June 02). "We are very worried about Covid-19 spreading to great apes". Retrieved from https://www.heidi.news/geneva-solutions/we-are-very-worried-about-covid-19-spreading-to-great-apes-iucn-geneva
4. FutureLearn. (2020, May 21). What Connects Deforestation and COVID-19? Retrieved from https://www.futurelearn.com/info/blog/deforestation-and-covid-19
5. Troëng, S., Barbier, E., & Rodríguez, C. (2020, May 21). COVID-19 is not a break for nature – let's make sure there is one after the crisis. Retrieved from https://www.weforum.org/agenda/2020/05/covid-19-coronavirus-pandemic-nature-environment-green-stimulus-biodiversity/

Chapter 5

1. Alberta Health Services. (2020, August 14). Novel coronavirus (COVID-19) Information for AHS Staff & Health Professionals. Retrieved from https://www.albertahealthservices.ca/topics/Page16947.aspx
2. Ambrose, J. (2020, April 30). Covid-19 crisis will wipe out demand for fossil fuels, says IEA. Retrieved from https://www.theguardian.com/business/2020/apr/30/covid-19-crisis-demand-fos-

sil-fuels-iea-renewable-electricity
3. Holden, E. (2020, July 07). Over 5,600 fossil fuel companies have taken at least $3bn in US Covid-19 aid. Retrieved from https://www.theguardian.com/environment/2020/jul/07/fossil-fuel-industry-coronavirus-aid-us-analysis
4. Normile, D. (2020, July 03). Novel human virus? Pneumonia cases linked to the seafood market in China stir concern. Retrieved from https://www.sciencemag.org/news/2020/01/novel-human-virus-pneumonia-cases-linked-seafood-market-china-stir-concern
5. Turk, D., & Kamiya, G. (2020, June 11). The impact of the Covid-19 crisis on clean energy progress – Analysis. Retrieved from https://www.iea.org/articles/the-impact-of-the-covid-19-crisis-on-clean-energy-progress
6. WHO. (2020, March 29). Modes of transmission of virus causing COVID-19: Implications for IPC precaution recommendations. Retrieved from https://www.who.int/news-room/commentaries/detail/modes-of-transmission-of-virus-causing-covid-19-implications-for-ipc-precaution-recommendations

Chapter 6

1. Baird, C. (2014, September 23). Does wasting household water remove it from the water cycle? Retrieved August 14, 2020, from https://wtamu.edu/~cbaird/sq/2014/09/23/does-wasting-household-water-remove-it-from-the-water-cycle/
2. Canada, H. (2016, March 23). Government of Canada. Retrieved August 14, 2020, from https://www.canada.ca/en/health-canada/services/air-quality/health-effects-indoor-air-pollution.html
3. Drop, A. (2018, April 12). Why We Should Be More Concerned About Wasting Water. Retrieved August 14, 2020, from https://dropconnect.com/blog/why-we-should-be-more-concerned-about-wasting-water
4. Edmond, C. (2020, June 11). How face masks, gloves and other coronavirus waste is polluting our ocean. Retrieved August 15, 2020, from https://www.weforum.org/agenda/2020/06/ppe-masks-gloves-coronavirus-ocean-pollution/
5. Environmental Defense Fund (2020). Health impacts of air pollution. Retrieved August 14, 2020, from https://www.edf.org/health/health-impacts-air-pollution
6. Feldman, D., Changoiwala, P., & Mackelprang, B. (2020, March 23). Amid Coronavirus Pandemic, Billions Lack Access to Clean Water. Retrieved August 14, 2020, from https://undark.org/2020/03/24/coronavirus-clean-water/
7. Kienapple, B. (2020, June 14). Coronavirus's Impact on the Environment [Infographic]. Retrieved August 14, 2020, from https://venngage.com/blog/coronavirus-impact-on-environment-infographic/

8. Liebsch, T. (2020, July 08). The rise of the face mask: What's the environmental impact of 17 million N95 masks? Retrieved August 15, 2020, from https://ecochain.com/knowledge/footprint-face-masks-comparison/

9. Londoño, E., Andreoni, M., & Casado, L. (2020, June 06). Amazon Deforestation Soars as Pandemic Hobbles Enforcement. Retrieved August 14, 2020, from https://www.nytimes.com/2020/06/06/world/americas/amazon-deforestation-brazil.html

10. Mukhopadhyay, S. (2020, August 14). COVID-19: Unmasking the Environmental Impact: Earth.Org - Past: Present: Future. Retrieved August 15, 2020, from https://earth.org/covid-19-unmasking-the-environmental-impact/

11. OpenStax. (2014, May 2). Biology II. Retrieved August 14, 2020, from https://courses.lumenlearning.com/suny-biology2xmaster/chapter/the-importance-of-biodiversity-to-human-life/

12. Robbins, J. (2016, February 23). How Forest Loss Is Leading To a Rise in Human Disease. Retrieved August 15, 2020, from https://e360.yale.edu/features/how_forest_loss_is_leading_to_a_rise_in_human_disease_malaria_zika_climate_change

13. World Health Organization, W. (2012, December 03). Biodiversity. Retrieved August 14, 2020, from https://www.who.int/globalchange/ecosystems/biodiversity/en/

14. World Wildlife Fund, W. (2010, March 19). Human health linked directly to forest health. Retrieved August 15, 2020, from https://wwf.panda.org/?unewsid=191323

Chapter 7

1. Araujo, M., & Naimi, B. (2020). Spread of SARS-CoV-2 Coronavirus likely to be constrained by climate. doi: 10.1101/2020.03.12.20034728

2. Do weather conditions influence the transmission of the coronavirus (SARS-CoV-2)? - CEBM. (2020). Retrieved 19 July 2020, from https://www.cebm.net/covid-19/do-weather-conditions-influence-the-transmission-of-the-coronavirus-sars-cov-2/

3. Harvey, C. (2020). Summer Weather Won't Save Us from Coronavirus. Retrieved 19 July 2020, from https://www.scientificamerican.com/article/summer-weather-wont-save-us-from-coronavirus/

4. Jüni, P., Rothenbühler, M., Bobos, P., Thorpe, K., da Costa, B., & Fisman, D. et al. (2020). Impact of climate and public health interventions on the COVID-19 pandemic: a prospective cohort study. Canadian Medical Association Journal, 192(21), E566-E573. doi: 10.1503/cmaj.200920

5. Luo, W., Majumder, M., Liu, D., Poirier, C., Mandl, K., Lipsitch, M., & Santillana, M. (2020). The role of absolute humidity on transmission rates of the COVID-19 outbreak. doi:

10.1101/2020.02.12.20022467

6. Poirier, C., Luo, W., Majumder, M., Liu, D., Mandl, K., Mooring, T., & Santillana, M. (2020). The Role of Environmental Factors on Transmission Rates of the COVID-19 Outbreak: An Initial Assessment in Two Spatial Scales. SSRN Electronic Journal. doi: 10.2139/ssrn.3552677

7. Sajadi, M., Habibzadeh, P., Vintzileos, A., Shokouhi, S., Miralles-Wilhelm, F., & Amoroso, A. (2020). Temperature, Humidity, and Latitude Analysis to Estimate Potential Spread and Seasonality of Coronavirus Disease 2019 (COVID-19). JAMA Network Open, 3(6), e2011834. doi: 10.1001/jamanetworkopen.2020.11834

8. The relationship between coronavirus (COVID-19) spread and the weather - Berkeley Earth. (2020). Retrieved 19 July 2020, from http://berkeleyearth.org/archive/coronavirus-and-the-weather/

9. Wang, M., Jiang, A., Gong, L., Luo, L., Guo, W., & Li, C. et al. (2020). Temperature significant change COVID-19 Transmission in 429 cities. doi: 10.1101/2020.02.22.20025791

10. WFP - COVID-19 and climate: Possible geographical and temporal patterns | Food Security Cluster. (2020). Retrieved 19 July 2020, from https://fscluster.org/coronavirus/document/wfp-covid-19-and-climate-possible

11. Yuan, S., Jiang, S., & Li, Z. (2020). Do Humidity and Temperature Impact the Spread of the Novel Coronavirus?. Frontiers In Public Health, 8. doi: 10.3389/fpubh.2020.00240

Chapter 8

1. Budds, D. (2020, March 17). Design in the age of pandemics. Retrieved August 18, 2020, from https://www.curbed.com/2020/3/17/21178962/design-pandemics-coronavirus-quarantine

2. Crosbie, M. (2020, July 05). How Might the COVID-19 Change Architecture and Urban Design? Retrieved August 18, 2020, from https://commonedge.org/how-might-the-covid-19-pandemic-change-architecture-and-urban-design/

3. Crosbie, M. (2020, March 08). When It Comes to Covid-19, Density Doesn't Kill-Sprawl Does. Retrieved August 18, 2020, from https://commonedge.org/when-it-comes-to-covid-19-density-doesnt-kill-sprawl-does/

4. Giacobbe, A. (2020, March 18). How the COVID-19 Pandemic Will Change the Built Environment. Retrieved August 18, 2020, from https://www.architecturaldigest.com/story/covid-19-design

5. Makhno, S. (2020, March 25). Life after coronavirus: How will the pandemic affect our homes? Retrieved August 18, 2020, from https://www.dezeen.com/2020/03/25/life-after-coronavirus-impact-homes-design-architecture/

6. Shima Hamidi, Sadegh Sabouri & Reid Ewing (2020) Does Density Aggravate the COVID-19 Pandemic?, Journal of the Ameri-

can Planning Association, DOI: 10.1080/01944363.2020.1777891
7. Wainwright, O. (2020, April 13). Smart lifts, lonely workers, no towers or tourists: Architecture after coronavirus. Retrieved August 18, 2020, from https://www.theguardian.com/artanddesign/2020/apr/13/smart-lifts-lonely-workers-no-towers-architecture-after-covid-19-coronavirus

Chapter 9

1. The Amazon Basin Forest | Global Forest Atlas. (n.d.). Yale University. Retrieved July 15, 2020, from https://globalforestatlas.yale.edu/region/amazon#:%7E:text=Beginning%20in%20the%201907%27s%20and,Amazon%20forest%20has%20been%20cleared
2. Bobylev, S. N. (2020). Environmental consequences of COVID-19 on the global and Russian economics. Population and Economics, 4(2), 43–48. https://doi.org/10.3897/popecon.4.e53279
3. Gardener, B. (2020, June 18). Why COVID-19 will end up harming the environment. National Geographic. https://www.nationalgeographic.com/science/2020/06/why-covid-19-will-end-up-harming-the-environment/
4. Y. (2020, April 3). The Unexpected Environmental Consequences of COVID-19. Voices of Youth. https://www.voicesofyouth.org/blog/unexpected-environmental-consequences-covid-19
5. Zambrano-Monserrate, M. A., Ruano, M. A., & Sanchez-Alcalde, L. (2020). Indirect effects of COVID-19 on the environment. Science of The Total Environment, 728(1), 138813. https://doi.org/10.1016/j.scitotenv.2020.138813

Chapter 10

1. Darrell Romuld. (2020, May 25). #JustCurious Has power and water use increased amid COVID-19? CTV News. https://regina.ctvnews.ca/just-curious/justcurious-has-power-and-water-use-increased-amid-covid-19-1.4954103
2. Drinking water management during the COVID-19 pandemic. (n.d.). Gouvernment Du Quebec. https://www.quebec.ca/en/environment-and-natural-resources/covid-19-environnement/drinking-water-management-covid-19/
3. Edmond, C. (2020, June 11). How face masks, gloves and other coronavirus waste is polluting our ocean. World Economic Forum. https://www.weforum.org/agenda/2020/06/ppe-masks-gloves-coronavirus-ocean-pollution/
4. Fadare, O. O., & Okoffo, E. D. (2020). Covid-19 face masks: A potential source of microplastic fibers in the environment. Science of The Total Environment, 737(1), 140279. https://doi.org/10.1016/j.scitotenv.2020.140279
5. Feng, E. (2020, March 16). NPR Choice page. NPR. https://

choice.npr.org/index.html?origin=https://www.npr.org/sections/goatsandsoda/2020/03/16/814929294/covid-19-has-caused-a-shortage-of-face-masks-but-theyre-surprisingly-hard-to-make

6. Gardener, B. (2020, June 18). Why COVID-19 will end up harming the environment. National Geographic. https://www.nationalgeographic.com/science/2020/06/why-covid-19-will-end-up-harming-the-environment/

7. Lassman, A. (2020, May 13). Impacts of Water Usage During Coronavirus Pandemic. NBC 6 South Florida. https://www.nbcmiami.com/news/local/changing-climate-south-florida/impacts-of-water-usage-during-coronavirus-pandemic/2233103/

8. Liebsch, T. (2020, July 8). The rise of the face mask: What's the environmental impact of 17 million N95 masks? Ecochain. https://ecochain.com/knowledge/footprint-face-masks-comparison/

9. McCandless, M. (2020, June 1). McCandless: COVID-19 has changed our fresh-water use. We need to be careful. Ottawa Citizen. https://ottawacitizen.com/opinion/mccandless-covid-19-has-changed-our-fresh-water-use-we-need-to-be-careful

10. Mendoza, N. F. (2020, June 2). US home water use up 21% daily during COVID-19 crisis. TechRepublic. https://www.techrepublic.com/article/us-home-water-use-up-21-daily-during-covid-19-crisis/

11. Monella, L. M. (2020, May 13). Will plastic pollution get worse after the COVID-19 pandemic? Euronews. https://www.euronews.com/2020/05/12/will-plastic-pollution-get-worse-after-the-covid-19-pandemic

12. Mukhopadhyay, S. (2020, August 14). COVID-19: Unmasking the Environmental Impact. Earth.Org - Past | Present | Future. https://earth.org/covid-19-unmasking-the-environmental-impact/

13. N95 Masks and the Coronavirus: More Production Underway. (n.d.). Honeywell. Retrieved July 12, 2020, from https://www.honeywell.com/en-us/newsroom/news/2020/03/n95-mask-and-the-coronavirus-more-production-underway

14. The face mask global value chain in the COVID-19 outbreak: Evidence and policy lessons. (2020, May 4). Organisation for Economic Co-Operation and Development (OECD). http://www.oecd.org/coronavirus/policy-responses/the-face-mask-global-value-chain-in-the-covid-19-outbreak-evidence-and-policy-lessons-a4df866d/

Chapter 11

1. Gibbens, S. (2020, April 21). Will the sustainable travel movement survive coronavirus? National Geographic. https://www.nationalgeographic.com/travel/2020/04/will-sustainable-travel-survive-coronavirus/
2. Hodd, C. (2020, June 5). COVID-19's environmental impact not likely sustainable: Climate experts say emissions down 8 per cent. Saltwire Network. https://www.saltwire.com/news/provincial/covid-19s-environmental-impact-not-likely-sustainable-climate-experts-say-emissions-down-8-per-cent-458523/
3. Mandal, I., & Pal, S. (2020). COVID-19 pandemic persuaded lockdown effects on environment over stone quarrying and crushing areas. Science of The Total Environment, 732(1), 139281. https://doi.org/10.1016/j.scitotenv.2020.139281
4. Milman, O. (2020, July 1). Pandemic side-effects offer glimpse of alternative future on Earth Day 2020. The Guardian. https://www.theguardian.com/environment/2020/apr/22/environment-pandemic-side-effects-earth-day-coronavirus
5. Newburger, E. (2020, March 22). Air pollution falls as coronavirus slows travel, but scientists warn of longer-term threat to climate change progress. CNBC. https://www.cnbc.com/2020/03/21/air-pollution-falls-as-coronavirus-slows-travel-but-it-forms-a-new-threat.html
6. Wikipedia contributors. (2020, August 15). Impact of the COVID-19 pandemic on the environment. Wikipedia. https://en.wikipedia.org/wiki/Impact_of_the_COVID-19_pandemic_on_the_environment

Chapter 12

1. Field, H. (2020, April 29). COVID-19 will herald an automation boom. Protocol. https://www.protocol.com/automation-boom-caused-by-coronavirus
2. Lazenby, H. (2020, April 8). Industry could fast-track automation amid COVID-19 fallout. Mining Journal. https://www.mining-journal.com/covid-19/news/1384090/industry-could-fast-track-automation-amid-covid-19-fallout
3. Necula, D., Vasile, N., & Stan, M. F. (2013). The impact of the electrical machines on the environment. 2013 8TH INTERNATIONAL SYMPOSIUM ON ADVANCED TOPICS IN ELECTRICAL ENGINEERING (ATEE), 1(1), 1–4. https://doi.org/10.1109/atee.2013.6563397
4. PricewaterhouseCoopers. (n.d.). COVID-19: What it means for industrial manufacturing. PwC. Retrieved August 8, 2020, from https://www.pwc.com/us/en/library/covid-19/coronavirus-impacts-industrial-manufacturing.html

5. Seric, A., & Winkler, D. (2020, April 22). Managing COVID-19: Could the coronavirus spur automation and reverse globalization? Industrial Analytics Platform. https://iap.unido.org/articles/managing-covid-19-could-coronavirus-spur-automation-and-reverse-globalization

6. Wang, L. (2020, July 8). Post–COVID-19: China's machinery production and automation equipment market cannot be immune to the global pandemic - Omdia. OMDIA Technology. https://technology.informa.com/624995/postcovid-19-chinas-machinery-production-and-automation-equipment-market-cannot-be-immune-to-the-global-pandemic

Chapter 13

1. Danton Unger. (2020, May 2). The five ways Canadians may change in the post-pandemic world. CTV News. https://winnipeg.ctvnews.ca/the-five-ways-canadians-may-change-in-the-post-pandemic-world-1.4922001

2. Dartnell, L. (2020, June 29). The Covid-19 changes that could last long-term. BBC Future. https://www.bbc.com/future/article/20200629-which-lockdown-changes-are-here-to-stay

3. Das, P. (2020, April 20). Lifestyle changes during Covid-19. Times of India Blog. https://timesofindia.indiatimes.com/blogs/melange/lifestyle-changes-during-covid-19/

4. Di Renzo, L., Gualtieri, P., Pivari, F., Soldati, L., Attinà, A., Cinelli, G., Leggeri, C., Caparello, G., Barrea, L., Scerbo, F., Esposito, E., & De Lorenzo, A. (2020). Eating habits and lifestyle changes during COVID-19 lockdown: an Italian survey. Journal of Translational Medicine, 18(1), 1–5. https://doi.org/10.1186/s12967-020-02399-5

5. Kunst, A. (2020, July 3). COVID-19 pandemic - Changes to general lifestyle 2020. Statista. https://www.statista.com/statistics/1105960/changes-to-the-general-lifestyle-due-to-covid-19-in-selected-countries/

6. Mahalingam, M. (2020, May 20). Let's Retain Pandemic-induced Healthy Lifestyle Changes, For They Will Help Build A Better World. Outlook India. https://www.outlookindia.com/website/story/opinion-learning-to-live-after-not-just-with-covid-19-should-be-the-way-forward/353162

7. Pearson, B. (2020, May 26). Humble Home To Power House: 6 Covid-19 Lifestyle Changes Retailers Should Consider. Forbes. https://www.forbes.com/sites/bryanpearson/2020/05/26/humble-home-to-power-house-6-covid-19-lifestyle-changes-retailers-should-consider/#42d9f50f38fb

8. Svoboda, E. (2020, April 17). How The COVID-19 Pandemic Will Change the Way We Live. Discover Magazine. https://www.discovermagazine.com/health/how-the-covid-19-pandemic-will-change-the-way-we-live

Chapter 14

1. Bruckner, M., & Mollerus, R. (2020, May 1). COVID-19 and the least developed countries | Department of Economic and Social Affairs. Retrieved August 16, 2020, from https://www.un.org/development/desa/dpad/publication/un-desa-policy-brief-66-covid-19-and-the-least-developed-countries/
2. Dettmer, J. (2020, June 12). Coronavirus Spreading Fast in Developing World. Retrieved August 16, 2020, from https://www.voanews.com/covid-19-pandemic/coronavirus-spreading-fast-developing-world
3. Friedman, S. (2020, July 15). COVID-19 has blown away the myth about 'First' and 'Third' world competence. Retrieved August 16, 2020, from https://theconversation.com/covid-19-has-blown-away-the-myth-about-first-and-third-world-competence-138464
4. Slaughter, G. (2020, April 29). 34 vulnerable countries could see 1B cases of COVID-19, prolonging pandemic. Retrieved August 16, 2020, from https://www.ctvnews.ca/health/coronavirus/34-vulnerable-countries-could-see-1b-cases-of-covid-19-prolonging-pandemic-1.4916388

www.ingramcontent.com/pod-product-compliance
Lightning Source LLC
Chambersburg PA
CBHW022109160426
43198CB00008B/408